Interactive Linear Algebra with MAPLE V®

Springer
New York
Berlin
Heidelberg
Barcelona
Budapest
Hong Kong
London
Milan
Paris
Santa Clara
Singapore
Tokyo

Interactive Linear Algebra with MAPLE V®

Elias Deeba

Ananda Gunawardena

Department of Computer and Mathematical Sciences
The University of Houston–Downtown

Springer

Textbooks in Mathematical Sciences

Series Editors:

Thomas F. Banchoff
Brown University

Jerrold Marsden
California Institute of Technology

Keith Devlin
St. Mary's College

Stan Wagon
Macalester College

Gaston Gonnet
ETH Zentrum, Zürich

Library of Congress Cataloging-in-Publication Data
Deeba, Elias Y.
    Interactive linear algebra with Maple V / Elias Y. Deeba, Ananda
D. Gunawardena.
        p.   cm. — (Textbooks in mathematical sciences)
    Includes bibliographical references and index.
    ISBN 0-387-98240-X (softcover : alk. paper).
    1. Algebras, Linear—Computer-assisted instruction.   2. Maple
(Computer file)   I. Gunawardena, Ananda D.   II. Title.
III. Series.
QA185.C65D44   1997
512´.5´078553042—dc21   97-26974

Printed on acid-free paper.

Production managed by Victoria Evarretta; manufacturing supervised by Jacqui Ashri.
Photocomposed copy prepared with the authors' LaTeX files by Bartlett Press, Marietta, GA.
Printed and bound by Hamilton Printing Co., Rensselaer, NY.
Printed in the United States of America.

9 8 7 6 5 4 3 2 1

ISBN 0-387-98240-X   Springer-Verlag   New York   Berlin   Heidelberg   SPIN 10523783

Contents

Preface

In our attempt to address issues pertaining to incorporating technology in the teaching of mathematics and issues of curriculum reform, we have focused on developing an introductory course in linear algebra. During all the stages of development, we tried to implement a philosophy based on our fundamental belief that students who study mathematics must enjoy, understand, assimilate, and apply the skills and concepts they study. Studying mathematics should be a worthwhile experience rather than a frustrating one. Students using this text learn the concepts and applications of linear algebra in an interactive environment characterized by experimentation, exploration and discovery learning. Students play a fundamental and active role rather than a passive one throughout the learning process.

Following are the guiding factors that influenced us throughout the development of the text:

- Linear algebra is an ideal course in which to blend theory, application, and computation.

- Applications of linear algebra are diverse and accessible.

- The concepts and applications of linear algebra can be learned in an active environment involving experimentation, exploration and discovery.

- Application problems from various disciplines such as computer science, graph theory, natural sciences, business economics, and population dynamics make the subject more interesting.

- The computational aspects of linear algebra are pertinent to emerging disciplines such as computational sciences.

- A unifying theme (solvability of linear equations) ties the units of the course together.

- Computer algebra systems (CASs) such as MapleV can be used effectively to develop an interactive self-paced electronic tutoring system.

- Effective use of technology enhances the understanding of linear algebra through graphical representations and animations.

- Labs and applications provide students with a variety of learning schemes that encourage collaborative learning, experimentation, and discovery.

- An adaptive testing system gives the students an accurate measure of their progress.

- An informal presentation style with less emphasis on abstraction may improve retention rate.

CONTENT OF THE TEXT

The interactive text consists of six units:

- One   Systems of Linear Equations
- Two   Algebra of Matrices
- Three   Linear Spaces
- Four   Inner Product Spaces
- Five   Linear Transformations
- Six   Eigenspaces

We include material that is normally excluded from an introductory linear algebra course. For example, LU-decomposition is included in Unit Two, least squares, generalized inverses, and QR-decomposition are included in Unit Three, and SVD-decomposition is included in Unit Six.

COMPONENTS OF THE TEXT

The main components of each unit in *Interactive Linear Algebra with Maple V* are library of automated functions, interactive computer lessons, supervised laboratories, applications, and theory, accompanied by a stand alone adaptive testing system.

The components are supported by a Graphical User Interface (GUI) designed to allow the user to navigate smoothly through the various components of the text. The main screen of the interface provides access to the six units of the text, to the index of automated functions, to the table of contents and the general help. Each unit screen includes an index of the material that can be accessed from that screen. For example, once a unit is selected, the user may click on either lessons, labs, applications, theory, or automated functions. Screens of lessons, labs, and applications include animations of linear algebra concepts. The GUI provides immediate access to the CAS Maple V.

Library of Automated Linear Algebra Functions

An automated function is a procedure written using a computer algebra system (CAS) capable of processing both numerical and symbolic inputs and whose output may include numerical, textual, and/or graphical components. The functions in *Interactive Linear Algebra with Maple V* represent an automation of the various concepts of the course. The interface to the functions is carefully designed and standardized to be intuitive and simple to use. The functions eliminate the emphasis on learning the syntax of the CAS and allow the user to focus and spend more time learning the ideas of the course. Also as a result of repeated use of the functions, students learn the relevant commands and the syntax of the CAS.

For consistency, the automated functions in our library have, for the most part, the same name as the Linalg package of Maple V. The functions serve in three capacities:

1. They are a **learning tool**: The **demonstration mode** shows the intermediate steps needed to learn a particular concept or algorithm.

2. They are a **testing tool**: The **interactive mode** allows the student to respond to a set of specific questions interactively. The questions are designed to ensure that the main features of an algorithm or a concept are highlighted. Such functions not only overcome laborious computations but enhance student learning of the process and the understanding of concepts. This encourages meaningful, active participation and experimentation in the learning process.

3. They can be **shortcuts**: The **nostep mode**, yields the desired output with no intermediate steps.

Each automated function has online help that includes a review of the concept or algorithm, examples, and the calling sequence of the function. Several other user-friendly options and features, such as error recovery and undo, are also included. Included in the software is a collection of demos that show how to use the automated functions. Students can go through the demonstration mode of these functions, in a self-paced manner, to learn the details of the underlying concept and then go through the interactive mode to check whether they have learned the concept. The demonstration mode and the interactive mode act to some extent as an "intelligent electronic tutor". The automated functions can be accessed from the main menu or from each unit menu. The main menu includes an index for all the functions. Each unit menu includes the automated functions relevant to the unit. Examples regarding the features and use of the functions can be accessed either from the main menu or from each unit menu. The following packages are included:

- **linsys**: `gausselim, rref, backsub, solveqns, graph`
- **linmat:** `LUdecomp, inverse, trainnet, commute, trsum, trproduct, trinverse, transtrans, matrixmul, Geometry`
- **linspace:** `lincomb, lindep, basis, subspace, graphvectadd, graphscalarmulti, graphlincomb`
- **linpdt:** `GramSchmidt, QRdecomp, leastsqrs, lsqrdemo`
- **lintran:** `lineartran, matrixrep, kernel, range, changebasis, BasisGeometry`
- **lineign:** `eigenvals, eigenvects, diagonalize, SVdecomp, hermitian, evplot`

Interactive Computer Lessons

Each lesson includes, objectives, motivations, examples, questions, exercises, a quiz, and a summary of the concepts and facts discussed. The lessons are informally developed and do not include formal proofs. Proofs of the facts can be accessed from the Theory component of the GUI. Most lessons include a Learning the Process section where students choose their favorite examples to understand a procedure. Students can develop, to some extent, their own notebooks. Each lesson is carefully designed to incorporate the relevant automated linear algebra functions. The interactive computer lessons allow students self-pacing and electronic tutoring.

Each lesson includes a collection of exercises. Solutions to the exercises are also provided in the software.

Each lesson includes a quiz. At the end of the quiz a report is provided that includes the student's score, number of correct answers, and questions that have not been answered correctly.

Supervised Labs

The purpose of the labs is to reinforce the concepts and skills learned in the lessons. Each lab includes a main task or two and one or two extra challenging problems for further exploration. Numerical aspects are emphasized. The automated functions are needed to work through the labs.

Applications

The applications demonstrate the utility of the concepts and algorithms studied in the lessons and the labs. The applications are modeling problems from various disciplines (such as difference and differential equations, natural sciences, computer science, networks, economics, population dynamics, and engineering).

Theory

The theory component includes proofs of linear algebra facts. We recommend that instructors select specific proofs and encourage students to go over these proofs. In MapleV Release 4, proofs are also hyperlinked to the lessons.

Adaptive Testing System

The adaptive testing system is a stand alone system that accompanies the Interactive Linear Algebra with Maple V. It includes a student package as well as an instrutor's package. The user has the option of selecting either an adaptive or nonadaptive form of the test. The test interface includes a calculator, access to MapleV, hints, explanation, confirm button, and access to general help. The purpose of the adaptive version is to provide immediate feedback on students' progress in learning a particular concept or algorithm. The degree of difficulty of the questions in the adaptive version of the test is based on the number of correct answers. The nonadaptive version provides randomly selected questions. In the nonadaptive form, the user must specify the number of questions to be attempted. There are over 400 multiple-choice questions each ranked according to degree of difficulty.

At the end of the test a report is provided that includes the student's score, number of correct answers, questions that have not been attempted, and an explanation for the incorrectly answered questions. The instructor's package included with the testing allows instructors to add new questions, modify existing ones, and keep a detailed record of the students. The password to access the instructor's package is bgd.

SUGGESTIONS FOR IMPLEMENTATION

Interactive Linear Algebra with Maple V is available in the form of a CD-ROM accompanied by a paperbound textbook. The text is an edited version of the interactive lessons, supervised labs, applications, and theory. We recommend that, in the first meeting, the instructor familiarize the students with the importance of using technology in learning mathematics: what is involved in the process of using technology in education, the time factor, the method of evaluation and grading, and the rewards and benefits of their efforts. The **interactive lessons** can be used by the instructor to highlight the main ideas of the lesson or by the students during class time with instructor supervision or outside the class time at the students' own pace. **Supervised laboratories** are to be completed by every student following the lessons in a unit. The labs are designed to be completed under the instructor's supervision within one class period. We recommend the completion of at least six labs. **Applications** are team activities (two students per team) are to be completed outside the class time (1–2 weeks). We recommend the completion of at least four applications. Frequent adaptive Tests are recommended to monitor students, progress and reevaluate the method of instruction.

ACKNOWLEDGMENTS

The authors are highly indebted to their families for their continuous encouragement, patience, support, and understanding throughout the preparation and development stages of this text. We are indebted to our parents, who gave us light and guidance. We appreciate the comments, suggestions, and patience of the students who went over preliminary versions of the text. We thank a very special student, Carlos Uribe, for his work on the interface and the exchange of ideas. We also appreciate the time and effort of another student, Binh Phan, for developing the adaptive testing system. We thank our colleague Chris Birchak in the English Department for editing preliminary version of the text. Last but not least, we would like to thank Jerry Lyons, at Springer Verlag-New York, Vicky Evarretta, Karen Phillips, and Ken Dreyhaupt, the Department of Computer and Mathematical Sciences at the University of Houston–Downtown, and the Interactive Mathematics Text Project (IMTP) of the Mathematical Association of America for their support.

The authors would appreciate any suggestions, comments, or corrections that users of this interactive text may wish to communicate to us.

Elias Deeba (Deeba@dt.uh.edu)
Ananda Gunawardena (Guna@pitt.edu)

Systems of Linear Equations

Systems of linear equations arise in the mathematical modeling of numerous problems in disciplines such as applied mathematics, physical and social sciences, and engineering. The linear problem $Ax = b$ is a basis of many mathematical models. In this course, A represents a matrix. In other settings, A may represent a differential, an integral transformation, or a combination of both. The main idea is to find conditions under which the system of linear equations $Ax = b$ is consistent or not. What algorithms can be used to decide a consistency of the system? If the system is consistent, what algorithms can be applied to solve the system? If the system is not consistent, can we obtain an approximate solution to the system?

Examples of Systems of Linear Equations

In this lesson we shall consider **systems of linear equations** and analyze their **solution set**. Systems of linear equations arise in the mathematical modeling of many problems.

Initialize the packages

```
> with(linalg):with(linsys);
```

SYSTEMS OF LINEAR EQUATIONS

Let us begin with some examples of two equations with two unknowns.

EXAMPLE 1.1   Consider the system of two equations with two unknowns

```
> eq1:= x+y =2; eq2:=x -2*y=4;
```

Let us check graphically whether this system has a solution

```
> graph(eq1,eq2);
```

In this example the system has a **unique solution**.

EXAMPLE 1.2   Consider the system of two equations with two unknowns

```
> eq1:= x+ 2*y =2; eq2:=2*x +4*y=4;
```

Let us check graphically whether this system has a solution

```
> graph(eq1,eq2);
```

In this example the system has **infinitely many solutions**.

EXAMPLE 1.3   Consider the system of two equations with two unknowns

```
> eq1:= x+2*y =2; eq2:=x+2* y=-2;
```

Let us check graphically whether this system has a solution

```
> graph(eq1,eq2);
```

In this example the system has **no solution**.

The same graphing technique can be applied for a system with three unknowns.

EXAMPLE 1.4 Consider the system of three equations with three unknowns

```
> eq1:= x+y+z =2; eq2:=x-2*y-z=4; eq3:=x-y-z=1;
```

Let us check graphically whether this system has a solution

```
> graph(eq1,eq2,eq3);
```

In this example the system has a **unique solution**.

EXAMPLE 1.5 Consider the system of three equations with three unknowns

```
> eq1:= x+y+z =2; eq2:=x+y+z=2; eq3:=x-y-z=1;
```

Let us check graphically whether this system has a solution

```
> graph(eq1,eq2,eq3);
```

In this example the system has **infinitely many solutions**.

EXAMPLE 1.6 Consider the system of three equations with three unknowns

```
> eq1:= x+y+z =2; eq2:=x+y+z=10;
```

Let us check graphically whether this system has a solution

```
> graph(eq1,eq2);
```

In this example the system has **no solution**.

   Simple word problems lead to systems of linear equations.

EXAMPLE 1.7   System of two equations in two unknowns. The amount of $10,000 is invested in two mutual funds M_1 and M_2. The yearly return of fund M_1 is 15% and the return of fund M_2 is 22%. The total return after one year is $2000. How much was invested in each fund?

Let x and y be the amounts, in thousands, invested in funds M_1 and M_2, respectively. The equations that model the above problem are

```
> eq1:=x+y=10;eq2:=15/100*x+22/100*y=2;
```

Are there values of x and y that satisfy both equations?
Check algebraically. Select the `solveqns` demonstration mode if you want to view the intermediate steps; otherwise select the nostep mode.

```
> solveqns({eq1,eq2},{x,y});
```

Example 1.7 represents a system of two equations in two variables that has a **unique solution.**

EXAMPLE 1.8   System of three equations in three unknowns. The amount of $10,000 is invested in three mutual funds M_1, M_2, and M_3. The return on the investment in the respective funds is 12%, 15%, and 22%, respectively. The total return is $2000. The investor's strategy is to invest twice as much in fund M_2 as in fund M_1. How much should be invested in each fund to achieve this return?

Let x, y, and z be the amounts, in thousands, invested in mutual funds M_1, M_2, and M_3, respectively. The mathematical equations that model this problem are

```
> eq1:=x+y+z=10;eq2:=12/100*x+15/100*y+22/100*z=2;
  eq3:=2*x-y=0;
```

Are there values of x, y, and z that satisfy both equations?
Check algebraically

```
> solveqns({eq1,eq2,eq3},{x,y,z});
```

Example 1.8 represents a system of three equations in three variables that has a **unique solution**.

EXAMPLE 1.9   System of two equations in three unknowns. The amount of $10,000 is invested in three mutual funds M_1, M_2, and M_3. The return on the investment in the respective funds is 12%, 15%, and 22%, respectively. The total return is $2000. How much should be invested in each fund to achieve this return?

Let x, y and z be the amounts, in thousands, invested in mutual funds M_1, M_2, and M_3, respectively. The mathematical equations that model this problem are

```
> eq1:=x+y+z=10;  eq2:=12/100*x+15/100*y+22/100*z=2;
```

Are there values of x, y and z that satisfy both equations?
Check algebraically

```
> solveqns({eq1,eq2},{x,y,z});
```

Compute the maximum possible amount that can be invested in fund M3. For the maximum investment in M_3, what are the amounts that can be invested in the other funds?

Example 1.9 represents a system of two equations in three unknowns with **many solutions.** Note that one of the variables assumes an arbitrary value. The other two variables can be solved in terms of this variable. The variable that assumes an arbitrary value is called a **free variable**.

EXAMPLE 1.10   System of one equation in four unknowns

```
> eq1:=2*x+3*y-7*z+2*w=0;
```

Check algebraically

```
> solveqns(eq1,{x,y,z,w});
```

Example 1.10 represents a system of one equation in four variables with **infinitely many solutions** and has **three free variables**.

EXAMPLE 1.11 System of three equations in three unknowns.

```
> eq1:=x+y-z=1;eq2:=x+2*y-7*z=4;eq3:=2*x+3*y-8*z=8;
```

Are there values of x, y, and z that satisfy both equations?
Check algebraically

```
> solveqns({eq1,eq2,eq3},{x,y,z});
```

Example 1.11 represents a system of three equations in three unknowns that has **no solution.**

Examples 1.1–1.11 show that one of the following must hold for a system of linear equations

- The system has a unique solution.
- The system has infinitely many solutions.
- The system has no solution.

In general, a system of m equations in n unknowns variables x_i $(i = 1, 2, \ldots, n)$ is a system of the form

$$a_{11}x_1 + a_{12}x_2 + a_{13}x_3 + \cdots + a_{1n}x_n = b_1$$

$$a_{21}x_1 + a_{22}x_2 + a_{23}x_3 + \cdots + a_{2n}x_n = b_2$$

$$a_{31}x_1 + a_{32}x_2 + a_{33}x_3 + \cdots + a_{3n}x_n = b_3$$

$$\vdots$$

$$a_{m1}x_1 + a_{m2}x_2 + a_{m3}x_3 + \cdots + a_{mn}x_n = b_m$$

where the a_{ij}'s are the coefficients of the unknown variables and the b's are known quantities. If the number of equations (m) is equal to the number of variables (n), the system is a square system. If all the b's are equal to zero, the system is called a **homogeneous system**; otherwise, it is called a **nonhomogeneous system**.

EXERCISES

In the following exercises you may need to use the automated functions graph and solveqns to answer the questions.

1. Consider the system of linear equations

$$2x + 3y = 5$$
$$x + ky = 15$$

Determine values of the parameter k so that the system has
a. a unique solution;
b. no solution.
Can this system have many solutions? Explain.

2. Does there exist a solution to the following systems of linear equations? Analyze and comment on the reasons why or why not a solution exists.

a. $2x - y + 3z = 1, x - 4y + 2z = 2, x - 11y + 3z = 5$
b. $2x - y + 3z = 1, x - 4y + 2z = 2, x - 11y + 3z = 6$
c. $2x - y + 3z = 1, x - 4y + 2z = 2, x - 11y + 4z = 7$

3. The system of linear equations

$$x - 2y = 2, \ x - y = 2, \ 4x - 8y = d$$

is assumed to have a unique solution for a certain value of d. By plotting several graphs with different choices of d, determine the condition on d that gives this unique solution.

4. Find all values of d (if any) such that the following system has no solution

$$x - 2y = 2, \ x - y = 2, \ 4x - 8y = d$$

Hint: Plot several graphs with different choices of d and determine the condition on d that yields this result.

5. Repeat Exercise 3 for the system

$$2x - y + 3z = 1, \ x - 4y + 2z = 2, \ x - 3y + z = d$$

6. The amount of $10,000 is invested in three mutual funds M_1, M_2, and M_3. The return on the investment in the respective funds is 12%, 15%, and 22%. The sum of the amounts invested in funds M_1 and M_2 is equal to the amount invested in fund M_3. The total return amounts to k dollars. What is the minimum possible return? What is the maximum possible return? How much should be invested in each fund to achieve these possible return values?

Consistent and Inconsistent Systems

The solution to a system of linear equations consists of the set of all values of the unknown variables that satisfy the system simultaneously. This set is called the **solution set**. For any system of linear equations, one of the following must hold:

- The solution set is a singleton.
- The solution set is infinite.
- The solution set is empty.

A system is **consistent** if its solution set is not empty. The solution set, in this case, is either a **singleton** or **infinite**. Accordingly, the system has a **unique** solution or **infinitely many solutions**. A system is **inconsistent** if its solution set is **empty**; equivalently, the system has **no solution**.

Initialize the packages

```
> with(linalg):with(linsys);
```

HOMOGENEOUS SYSTEMS

A **homogeneous system** of m equations with n unknowns is a system of the form:

$$a_{11} x_1 + a_{12} x_2 + a_{13} x_3 + \cdots + a_{1n} x_n = 0$$
$$a_{21} x_1 + a_{22} x_2 + a_{23} x_3 + \cdots + a_{2n} x_n = 0$$
$$a_{31} x_1 + a_{32} x_2 + a_{33} x_3 + \cdots + a_{3n} x_n = 0$$
$$\vdots$$
$$a_{m1} x_1 + a_{m2} x_2 + a_{m3} x_3 + \cdots + a_{mn} x_n = 0$$

Is a homogeneous system consistent or inconsistent?
The homogeneous system of linear equations is always consistent. The solution set of a homogeneous system is never empty. In fact, $x_1 = 0, x_2 = 0, \ldots, x_n = 0$ always satisfy such systems. This solution is called the **zero** or **trivial solution**.

EXAMPLE 2.1   Consider the system

```
> eq1:=x+y+2*z=0; eq2:=2*x+y-3*z=0;
```

What is the solution set of this system?

```
> solveqns({eq1,eq2},{x,y,z});
```

Example 2.1 represents a system with **infinitely many solutions** and with **one free variable.**

EXAMPLE 2.2   Consider the system

```
> eq1:=x-y-z+2*w=0; eq2:= x+y+4*z-w=0;
eq3:=7*x-6*y+7*w=0;
```

What is the solution set of this system?

```
> solveqns({eq1,eq2,eq3},{x,y,z,w});
```

Example 2.2 represents a system with **infinitely many solutions** and with **one free variable.**

EXAMPLE 2.3   Consider the system

```
> eq1:=x+2*y-5*z+6*w=0;
```

What is the solution set of this system?

```
> solveqns(eq1,{x,y,z,w});
```

This system has **infinitely many solutions** with **three free variables**.

EXAMPLE 2.4   Consider the system

```
> eq1:=x-2*y-z=0; eq2:=2*x-y+z=0; eq3:=x+3*y-5*z=0;
```

What is the solution set of this system?

```
> solveqns({eq1,eq2,eq3},{x,y,z});
```

This system has the **trivial solution** as its **unique solution**.

Examples 2.1–2.4 illustrate several scenarios for homogeneous systems. Are all homogeneous systems consistent? Can you describe their solution set?

If the number of equations is fewer than the number of unknowns (underdetermined system), the homogeneous system will have infinitely many solutions. If the number of equations is greater or equal to the number of unknowns (overdetermined system), the homogeneous system will have either a unique solution or infinitely many solutions. Thus,

A homogeneous system of linear equations is consistent. The solution set is either a singleton or infinite. (Fact 1.1)

NONHOMOGENEOUS SYSTEMS

Consider the **nonhomogeneous** system of m equations with n unknowns:

$$a_{11}\,x_1 + a_{12}\,x_2 + a_{13}\,x_3 + \cdots + a_{1n}\,x_n = b_1$$

$$a_{21}\,x_1 + a_{22}\,x_2 + a_{23}\,x_3 + \cdots + a_{2n}\,x_n = b_2$$

$$a_{31}\,x_1 + a_{32}\,x_2 + a_{33}\,x_3 + \cdots + a_{3n}\,x_n = b_3$$

$$\vdots$$

$$a_{m1}\,x_1 + a_{m2}\,x_2 + a_{m3}\,x_3 + \cdots + a_{mn}\,x_n = b_m$$

Is the nonhomogeneous system consistent or inconsistent?

EXAMPLE 2.5   Consider the system

```
> eq1:=x+y+2*z=1; eq2:=2*x+2*y-4*z=2;
```

What is the solution set of this system?

```
> solveqns({eq1,eq2},{x,y,z});
```

This system has infinitely many solutions with one free variable. The solution $[x, y, z]$ of the nonhomogeneous system can be expressed as $[x, y, z] = [1 - a, a, 0] = [1, 0, 0] + [-a, a, 0] = [1, 0, 0] + a * [-1, 1, 0]$ where a is an arbitrary real variable. The solution $[1, 0, 0]$ is the **particular** solution to the nonhomogeneous system, while $[-1, 1, 0]$ is a solution of the associated homogeneous system.

EXAMPLE 2.6   Consider the system

```
> eq1:=x+2*y-5*z+6*w=10;
```

What is the solution set of this system?

```
> solveqns(eq1,{x,y,z,w});
```

This system has **infinitely many solutions** with **three free variables**. The general solution $[x,y,z,w]$ of the nonhomogeneous system is

$$[x, y, z, w] = [10 - 2 * a + 5 * b - 6 * c, a, b, c]$$

$$= [10, 0, 0, 0] + [-2 * a, a, 0, 0] + [5 * b, 0, b, 0] + [-6 * c, 0, 0, c]$$

$$= [10, 0, 0, 0] + a * [-2, 1, 0, 0] + b * [5, 0, 1, 0] + c * [-6, 0, 0, 1]$$

The solution $[10,0,0,0]$ is the particular solution of the system; the values $[-2,1,0,0]$; $[5,0,1,0]$; and $[-6,0,0,1]$ are solutions to the associated homogeneous system.

EXAMPLE 2.7   Consider the system

```
> eq1:=x-y-z=1; eq2:= x+y+4*z=-1; eq3:=7*x-6*y-z=7;
```

What is the solution set of this system?

```
> solveqns({eq1,eq2,eq3},{x,y,z});
```

Example 2.7 represents a system with a **unique solution**.

EXAMPLE 2.8   Consider the system

```
> eq1:=x-y-z=1; eq2:= x+y+4*z=-1; eq3:=3*x+y+7*z=3;
```

What is the solution set of this system?

```
> solveqns({eq1,eq2,eq3},{x,y,z});
```

Example 2.8 represents a system with **no solution**.

Examples 2.5–2.8 illustrate several situations for nonhomogeneous systems of linear equations. Is every nonhomogeneous system of linear equations consistent? Can you describe the solution set? Can you describe the general solution of a nonhomogeneous system of linear equations whose solution set is infinite?

Every system of linear equations Ax=b has either no solution, exactly one solution, or infinitely many solutions. (Fact 1.2)

EXERCISES

In the following exercises you may use the automated functions `graph` and `solveqns`.

1. Determine whether the given systems of linear equations are consistent. If the system is consistent with infinitely many solutions, determine the number of the free variables and express the solution in terms of these variables.

    a. $x - y + z - w = 0$, $2x + 3y - z + 2w = 0$, $x + 2y - 7z + w = 1$;
    b. $x - y + z = 1$, $2x + y - 3z = 2$, $3x - 2z = 4$;
    c. $x - y + 2z = 0$, $x + 4y - 3z = 2$, $3x - y + 2z = -1$;

2. Give an example of

    a. an inconsistent nonhomogeneous system of linear equations with fewer equations than unknowns.
    b. a consistent nonhomogeneous system of linear equations with fewer equations than unknowns.

3. Explain why a homogeneous system with more equations than unknowns will have either a unique solution or infinitely many solutions.

4. Explain why

    a. the sum of solutions of a homogeneous system of linear equations is also a solution to the system.

b. the scalar multiple of a solution of a homogeneous system of linear equations is also a solution to the system.

5. Give an example of

   a. an inconsistent nonhomogeneous system of linear equations whose associated homogeneous system has infinitely many solutions.

   b. a consistent nonhomogeneous system of linear equations whose associated homogeneous system has infinitely many solutions. Write the general solution of the nonhomogeneous system.

   c. a consistent nonhomogeneous system of linear equations whose associated homogeneous system has a unique solution. How many solutions does the nonhomogeneous system have?

6. Determine the polynomial of second degree $p(x) = ax^2 + bx + c$ that passes through the points: $P(1, 2)$; $Q(-1, 3)$; and $R(-4, 5)$. (*Hint*: Write the equations to determine the coefficients a,b, and c)

7. Is there a polynomial of second degree $p(x) = ax^2 + bx + c$ that passes through the points: $P(1, 2)$; $Q(-1, 3)$; $R(-4, 5)$ and $S(0, 2)$? Justify your answer.

8. Is there a polynomial of second degree $p(x) = ax^2 + bx + c$ that passes through the points: $P(1, 2)$; $Q(-1, 3)$? If so how many polynomials can you find? Justify your answer.

LESSON 1.3 Equivalent Systems and Elementary Row Operations

In this lesson we will reduce a given system of linear equations to another system that is easier to manipulate. What types of operations are permissible to reduce a system of linear equations into another "equivalent" system?

Initialize the packages

```
> with(linalg): with(linsys);
```

EQUIVALENT SYSTEMS

EXAMPLE 3.1   Consider the following systems of linear equations. System S_1 consists of the equations

```
> eq1:= x+y+z=1; eq2:= 2*x+3*y-z=2; eq3:=x+y+2*z=3;
```

System S_2 consists of the equations

```
> eq11:= x+y+z=1; eq22:=y-3*z = 0; eq33:=z = 2;
```

The solution of system S_1 is (use the nostep mode of solveqns)

```
> solveqns({eq1,eq2,eq3},{x,y,z});
```

The solution of the system S_2 is

```
> solveqns({eq11,eq22,eq33},{x,y,z});
```

Systems S_1 and S_2 have the same solution. System S_2 is easier to solve than system S_1. Indeed, by substituting the value $z = 2$ into eq22 we find y, and then by substituting the values of z and y into eq11 we find x.

Systems of linear equations that have the same solution set are called equivalent systems.

ELEMENTARY ROW OPERATIONS

Example 3.1 suggests that it may be worth the effort to reduce a system to an equivalent one that is easier to handle. What operations are permissable to do so?

Will the solution set of an equation change upon multiplying it by a nonzero scalar?

EXAMPLE 3.2   Consider the equation

```
> eq1:= x+2*y=3;
```

and its multiple

```
> eq2:=3*x+6*y=9;
```

Check graphically whether the two equations have the same solution set

```
> graph(eq1,eq2);
```

The graph implies that the two lines overlap. Thus, they have the same solution set. Check algebraically

```
> solveqns(eq1,{x,y});
> solveqns(eq2,{x,y});
```

Multiplying an equation of a system of linear equations by a nonzero scalar yields an equivalent system.

Will the solution set of a system of linear equations change if two equations are interchanged?

EXAMPLE 3.3   Consider the system S_1

```
> eq1:=x+y+3*z=4; eq2:=x-y+z=0; eq3:=x-y-z=10;
```

Interchange equation 2 and equation 3 of S_1 to get new system S_2

```
> eq11:=x+y+3*z=4; eq22:=x-y-z=10; eq33:=x-y+z=0;
```

Are systems S_1 and S_2 equivalent? Check the solution set

```
> solveqns({eq1,eq2,eq3},{x,y,z});
> solveqns({eq11,eq22,eq33},{x,y,z});
```

Since they have the same solution, the two systems are equivalent.

Interchanging equations of a system of linear equations yields an equivalent system.

Will the solution set of a system of equations change if we replace an equation by the sum of that equation and a nonzero multiple of another equation?

EXAMPLE 3.4   Consider the system S_1

```
> eq1:=x-2*y=2;eq2:=x+3*y=1;
```

Obtain new system S_2 by multiplying eq1 of S_1 by 2 and adding it to eq2

```
> eq11:=x-2*y=2; eq22:=3*x-y=5;
```

Check graphically whether both systems have the same solution set.
The graph of the system S_1 is

```
> graph(eq1,eq2);
```

The graph of the system S_2 is

```
> graph(eq11,eq22);
```

What do you conclude from the graphs of the two systems? The graphs indicate that the two systems have the same solution set. What does this imply about the two systems?

Replacing an equation by adding it to a multiple of another equation of a system of linear equations yields an equivalent system.

Examples 3.1–3.4 show that the following operations yield equivalent systems of linear equations:

- Multiply one equation by a nonzero constant.
- Interchange two equations.
- Add a multiple of one equation to another equation.

These three operations are called **elementary row operations**.

EXERCISES

1. Identify the elementary row operations that convert system A to system B and verify graphically using the automated function graph that the two systems are equivalent.

   a. A

   ```
   > eq1:=x-3*y=1;eq2:=2*x-5*y=3;
   ```

   B

   ```
   > eq1:=x-3*y=1;eq2:=y=1;
   ```

   b. A

   ```
   > eq1:=x-4*y=1;eq2:=x-y=3;
   ```

   B

   ```
   > eq1:=x-4*y=1;eq2:=3*x-9*y=5;
   ```

   c. A

   ```
   > eq1:=x-3*y=1;eq2:=3*x-5*y=3;
   ```

   B

```
> eq1:=3*x-9*y=3;eq2:=3*x-5*y=3;
```

d. A

```
> eq1:=x-y=1;eq2:=2*x+2*y=5;
```

B

```
> eq1:=2*x+2*y=5;eq2:=x-y=1;
```

2. Show algebraically that the following pairs of systems are equivalent.

a. A

```
> eq1:=x-y+z=1;eq2:=2*x-3*y+z=0;eq3:=-x+2*y-2*z=-1;
```

B

```
> eq1:=x-y+z=1;eq2:=y+z=2;eq3:=z=1;
```

b. A

```
> eq1:=x-y+z=3;eq2:=x-3*y+z=0;
```

B

```
> eq1:=x-y+z=3;eq2:=y=3/2;
```

LESSON 1.4 Matrix Representation of Linear Systems

In solving a system of linear equations, one may suppress the unknown variables and define a **matrix** to represent the underlying system. In general, an $m \times n$ matrix may be defined as a rectangular array of numbers consisting of m rows and n columns. Matrices constitute a convenient and efficient notation for summarizing and tabulating information for many practical problems. For example, the sale of seven different brands in three different locations of a department store could be represented by the matrix

$$
\begin{bmatrix}
20 & 10 & 30 & 15 & 22 & 17 & 56 \\
34 & 53 & 42 & 67 & 33 & 78 & 36 \\
22 & 14 & 56 & 78 & 93 & 17 & 33
\end{bmatrix}
$$

For systems of linear equations, we associate two types of matrices:

- **Coefficient matrix**: A matrix whose rows consist of the coefficients of the variables.
- **Augmented matrix**: A matrix whose rows consist of the coefficients of the variables and whose last column consists of the values of the equations. The rows of an $m \times (n+1)$ augmented matrix represent nonhomogeneous system of m equations with n unknowns.

Initialize the packages

```
> with(linalg):with(linsys);
```

MATRIX REPRESENTATION

EXAMPLE 4.1 Consider the system of equations

```
> eq1:=2*x-3*y+z-w=0;  eq2:=x-y+z-5*w=0;
eq3:=x-y+z-w=9;  eq4:=x+6*y-9*z+w=21;
```

The augmented matrix associated with this system is

```
> AUG:=matrix([[2,-3,1,-1,0],[1,-1,1,-5,0],
```

```
    [1,-1,1,-1,9],  [1,6,-9,1,21]]);
```

EXAMPLE 4.2 Consider the system of equations

```
> eq1:=2*x-3*y+z=0; eq2:=7*x+11*y-9*z=2;
```

The augmented matrix associated with this system is the 2×4 matrix

```
> AUG:=matrix([[2,-3,1,0],[7,11,-9,2]]);
```

We can also establish the following association: for every augmented matrix, one can write the corresponding system of linear equations.

EXAMPLE 4.3 Consider the augmented matrix

```
> A:=matrix([[2,3,5,1],[1,0,2,5],[2,4,0,0]]);
```

The set of equations associated with this matrix is

```
> eq1:=2*x+3*y+5*z=1; eq2:=x+2*z=5; eq3:=2*x+4*y=0;
```

EXAMPLE 4.4 Consider the augmented matrix

```
> A:=matrix([[2,3,5,1,0],[-1,8,2,5,0],[2,4,9,-7,0]]);
```

The set of equations associated with this matrix is

```
> eq1:=2*x+3*y+5*z+w=0; eq2:=-x+8*y +2*z+5*w=0;
eq3:=2*x+4*y+9*z-7*w=0;
```

It is clear by now that matrix notation is a convenient way to represent a given system of linear equations.

EXERCISES

1. Given the augmented matrix of a system of linear equations

```
> A:=matrix([[1,2,-2,a],[-1,3,5,b],[0,5,3,c]]);
```

a. Write the nonhomogeneous system associated with the matrix A.
b. Does there exist a solution for any choice of the parameters a, b, and c?
  Is there any choice for a, b, and c so that the system has infinitely many solutions?

2. Given the system of linear equations

```
> eq1:=x-y+z-2*w=0; eq2:=2*x+3*y-4*z+w=1; ·
    eq3:=-3*x+6*y-3*z+w=2; eq4:=-x+y+6*z-k*w=2;
```

a. Write the coefficient matrix associated with the system.

b. Write the augmented matrix associated with the system.

c. Find the value of k for which the system is inconsistent. *Hint*: you may choose values of k ranging from -11 to -15. Within this range can you guess this value of k?

LESSON 1.5

Basic Linear Algebra Algorithms

The three elementary row operations discussed in Lesson 1.3 can be used to reduce the augmented matrix of a system of linear equations to a matrix from which we can deduce whether the system is consistent or not. If the system is consistent, how do we obtain the solution.

Initialize the following packages

```
> with(linalg):with(linsys);
```

REDUCED SYSTEMS

EXAMPLE 5.1   Assume we are given the system

```
> eq1:=x1+2*x2-3*x3=4; eq2:=-3*x1-6*x2-5*x3=7;
eq3:=-x1+5*x2-11*x3=12;
```

The augmented matrix associated with this system is

```
> AUG:=matrix([[1,2,-3,4],[-3,-6,-5,7],[-1,5,-11,12]]);
```

The idea is to get a system equivalent to the original one yet easier to solve.

We perform elementary row operations on AUG to eliminate x_1 from the second and third equation. The Maple commands for performing these operations are

- Multiply row 1 by 3 and add to row 2

  ```
  > AUG1:=addrow(AUG,1,2,3);
  ```

- Multiply row 1 by 1 and add to row 3

  ```
  > AUG2:=addrow(AUG1,1,3,1);
  ```

  We then eliminate x_2 from the third equation of the resulting system.

- Interchange row 2 and row 3

  ```
  > AUG3:=swaprow(AUG2,2,3);
  ```

- Multiply row 2 by 1/7

  ```
  > AUG4:=mulrow(AUG3,2,1/7);
  ```

- Multiply row 3 by −1/14:

  ```
  > AUG5:=mulrow(AUG4,3,-1/14);
  ```

The matrix AUG5 is referred to as the **echelon form** of AUG. The new system associated with AUG5 is

```
> eq1:=x1+2*x2-3*x3=4; eq2:=x2-2*x3=16/7; eq3:=x3=-19/14;
```

This system is equivalent to the original system since it is obtained by a sequence of elementary row operations.

How do we obtain the matrix AUG5 from the original augmented matrix AUG? To obtain the echelon form of the matrix AUG, we perform a sequence of elementary row operations until we get a form that resembles AUG5 in this example.

What does the reduced matrix AUG5 or the new system tell us about the solution of the original system? The last row of AUG5 (equivalently eq3 of the new system) indicates that the system is **consistent** and has a **unique solution**.

What changes can you propose to the original system so that the new system will have infinitely many solutions? You could change the (3,4)-entry 12 of the matrix AUG

```
> AUG:=matrix([[1,2,-3,4],[-3,-6,-5,7],[-1,5,-1 1,12]]);
```

to 15 and consider the new augmented matrix

```
> AUG:=matrix([[1,2,-3,4],[-3,-6,-5,7],[-1,-2,- 11,15]]);
```

Perform a sequence of elementary row operations on the new matrix AUG

```
> AUG1:=addrow(AUG,1,2,3);
> AUG2:=addrow(AUG1,1,3,1);
> AUG3:=mulrow(AUG2,2,-1/14);
> AUG4:=addrow(AUG3,2,3,14);
```

The matrix AUG4 is the echelon form of AUG. The new system associated with AUG4 is

```
> eq1:=x1+2*x2-3*x3=4; eq2:=x3=-19/14; eq3:=0=0;
```

The echelon form AUG4 of the matrix AUG indicates that this new system is **consistent** and has **infinitely many solutions**.

What changes can you propose to the original system so that the new system will have no solution? You could change the third row of the matrix AUG

```
> AUG:=matrix([[1,2,-3,4],[-3,-6,-5,7],[-1,5,-11,12]]);
```

and consider the new augmented matrix

```
> AUG:=matrix([[1,2,-3,4],[-3,-6,-5,7],[-2,-4,-8,15]]);
```

Perform a sequence of elementary row operations on the new matrix AUG

```
> AUG1:=addrow(AUG,1,2,3);
> AUG2:=addrow(AUG1,1,3,2);
> AUG3:=mulrow(AUG2,2,-1/14);
> AUG4:=addrow(AUG3,2,3,14);
```

The matrix AUG4 is the echelon form of AUG. The new system associated with AUG4 is

```
> eq1:=x1+2*x2-3*x3=4; eq2:=x3=-19/14; eq3:=0=4;
```

The echelon form AUG4 of the matrix AUG indicates that this new system is **inconsistent**.

We have used the elementary row operations to reduce a given matrix to its **Echelon form** to deduce the consistency of the associated system of linear equations. Following is a representation of an original system and its reduced form.

$$
\begin{array}{cc}
\textbf{Original system} & \textbf{Reduced system} \\
\begin{bmatrix}
* & * & * & * & * & * \\
* & * & * & * & * & * \\
* & * & * & * & * & * \\
* & * & * & * & * & * \\
* & * & * & * & * & * \\
* & * & * & * & * & *
\end{bmatrix}
\rightarrow
\begin{bmatrix}
1 & * & * & * & * & * \\
0 & 1 & * & * & * & * \\
0 & 0 & 1 & * & * & * \\
0 & 0 & 0 & 0 & 1 & * \\
0 & 0 & 0 & 0 & 0 & 1 \\
0 & 0 & 0 & 0 & 0 & 0
\end{bmatrix}
\end{array}
$$

BASIC ALGORITHMS

The reduced system should have the following properties

- The first non-zero entry in any row must be one. This is called a leading 1.
- All entries below the leading 1 must be made zero.
- The leading 1 in row $i \geq 2$ is always to the right of the leading 1 in previous rows.
- All zero rows must appear at the end of the matrix.

The resulting matrix is a matrix in row **echelon form.** In addition, if

- All entries above and below a leading 1 are zero.

then the resulting matrix is a matrix in **reduced row echelon form.** The algorithm that yields the row echelon form of a matrix is the **Gauss elimination algorithm**. The algorithm that yields the reduced row echelon form of a matrix is the **Gauss-Jordan algorithm**. If either form results in a consistent system of linear equations, then we apply **backsubstitution algorithm** to the reduced form of the augmented matrix to obtain the solution of the original system.

Learning the Process

Let us go over the Gauss Elimination Algorithm. Enter the matrix and then call the function `gausselim`. You can modify the text by choosing your favorite examples: You can experiment with any example you like.

EXAMPLE 5.2   Given the matrix

```
> A:=matrix([[1,2,-1],[2,4,6],[5,7,9]]);
```

choose the demonstration mode of `gausselim` function

```
> gausselim(A);
```

If the resulting matrix is the augmented matrix of a system of linear equations, is the system consistent or not? If consistent, how many solutions does it have?

EXAMPLE 5.3   Given the matrix

```
> A:=matrix([[5,2,-9,7],[2,4,6,11],[15,7,29,45] ]);
```

obtain the echelon form of the matrix

```
> gausselim(A);
```

If the resulting matrix is the augmented matrix of a system of linear equations, is the system consistent or not? If consistent, how many solutions does it have?

EXAMPLE 5.4   Given the matrix

```
> A:=matrix([[3,1,-9,6],[4,6,1,1],[7,7,-8,7]]);
```

the echelon form of A is obtained by

```
> gausselim(A);
```

If the resulting matrix is the augmented matrix of a system of linear equations, is the system consistent or not? If consistent, how many solutions does it have?

The reduced row echelon form of matrix A in Example 5.4 can be obtained by using the Guass-Jordan algorithm

```
> rref(A);
```

If the resulting matrix is the augmented matrix of a system of linear equations, is the system consistent or not? If consistent, how many solutions does it have?

Backsubtitution Algorithm

Let us now learn the backsubstitution algorithm.

EXAMPLE 5.5   Use backsub to solve the system whose augmented matrix is

> A:=matrix([[1,3,5,6,15],[2,3,7,9,11],[4,9,17, 21,41]]);

First put the matrix in row echelon form. Use the nostep mode of gausselim

> gausselim(A);

The echelon form of the matrix is

> A1:=matrix([[1,3,5,6,15],[0,1,1,1,19/3],[0,0, 0,0,0]]);

Apply the backsubstitution algorithm

> backsub(A1);

Repeat Examples 5.4 and 5.5 with matrices or systems of linear equations of your choice to learn these algorithms.

EXERCISES

In the following exercises you may need to use the automated functions gausselim, rref, and backsub.

1. The augmented matrix of a system of linear equations is given by

> A:=matrix([[1,-1,3,e],[11,3,5,f],[-12,32,k,g]]);

  a. Determine the values of k for which the system has a unique solution. Does the solution depend on the parameters e, f, and g?
  b. Is there a value of k for which the system has infinitely many solutions? If so, is this independent of the choice of the parameters e, f, and g, or does it depend on the values of these parameters?

2. Given the matrix

> A:=matrix([[3,-2,5,6],[-5,7,4,1],[-2,5,9,7]]);
  b:=matrix([[b1],[b2],[b3]]);

  a. Find all 3×1 matrices b for which there is at least one solution to the system $Ax = b$.
  b. Find all solutions associated with matrix b obtained in part (a).

3. Given the points $(1, 1.5), (2, 2.7), (3, 4.1)$ and $(4, 5.1)$, find the interpolating polynomial $p(x) = a + bx + cx^2 + dx^3$ that passes through these points. **Hint:** Obtain a system of equations in a, b, c, and d to determine the polynomial.

4. The coefficient matrix A of a homogeneous system of linear equations is

> A:=matrix([[1,1,3,5,7,2], [6,3,0,2,9,4], [3,1,6,8,9,6],
  [10,5,9,15,25,12]]);

a. Write the homogeneous system of linear equations.
b. How many solutions does this system have? What are the free variables? Describe the solution set.
c. Describe the solution set of the nonhomogeneous system of linear equations $Ax = b$ where b is the matrix:

```
> b:=matrix([[2],[4],[6],[12]]);
```

d. Repeat part (c) if the last entry of b is 10 instead of 12.

5. Given the system of linear equations

```
> eq1:=x+2xy-kyz=9; eq2:=2xy-y+3yz=6;eq3:=x+y+z=3;
```

a. Write down the augmented matrix associated with the system.
b. Apply an appropriate algorithm to infer conditions on parameter k so that the system has (1) a unique solution (2) no solution.
c. From the reduced form answer the following: (1) is there a value of k so that the system has a unique solution? (2) is there a value of k so that the system has no solution?

6. An investment firm has selected 5 mutual funds to invest part of their assets. The mutual funds range from low-risk funds to high-risk ones. The return of these funds follows.

M_1	M_2	M_3	M_4	M_5
10%	20%	16%	20%	24%

The amount of $1,000,000 is to be distributed among these funds to meet a projected return of $150,000 after one year. The following strategy is being applied: the sum of twice the amount invested in M_1 and the amounts invested in M_2 and M_5 should be equal to five times the amount invested in M_3; the sum of the amount invested in M_3 and four times the amount invested in M_4 should be the same as the sum of the amount invested in M_2 and M_5; the sum of the amount invested in M_1, M_4 and twice that of M_2 should be five times the amount invested in M_5.

a. Write the system of equations that models this problem.
b. Write the augmented matrix associated with the system.
c. How much should be invested in each fund to achieve the predicted return?

Graphical Representations

Purpose

The purpose of this lab is to determine graphically and algebraically whether a given system of linear equations is consistent. If it is consistent, use the backsubstitution algorithm to solve the system.

Automated Linalg Functions

In this lab, you will use the interactive mode of the automated functions: `gausselim`; `rref`; `backsub`. For help with function, type at the Maple prompt: `>?function name`; for example, `?gausselim`;

INSTRUCTIONS

1. To execute a statement, move the cursor to the line using the mouse or the keyboard and press the enter key.
2. While executing a function, create Maple input regions if needed; otherwise, the output will not appear in the desired place.
3. Execute the following commands to load the packages

```
>   with(linalg):with (linsys);
```

TASK 1

The purpose of this task is to check graphically whether a system of linear equations is consistent or inconsistent.

Activity 1

Consider the following system of linear equations

```
> eq1 := 2*x-4*y=5; eq2 := 13*x+7*y=4;
```

```
eq3 := 11*x+11*y=-1;
```

Graph the system using the procedure `graph`.

```
> graph(eq1,eq2,eq3);
```

Based on the graph, does the system have a unique solution, infinitely many solutions, or no solution? If there is a solution, estimate its value.

Activity 2

Change one equation of the system given in Activity 1 so that the new system will have no solution. Enter the new system of equations

```
> eq1 :=
> eq2 :=
> eq3 :=
```

Graph the new system using the function `graph`.

```
> graph(eq1,eq2,eq3);
```

Based on the graphs, is the system consistent or inconsistent?

Activity 3

Is it possible to change one equation of the system in Activity 1 so that the new system will have infinitely many solutions? Justify your answer.

Activity 4

Determine graphically, whether the following system is consistent

```
> eq1 :=2*x + 2*y + z = 2; eq2:= x + 2*y + z = 1;
> graph(eq1,eq2);
```

Activity 5

Add one equation to those of Activity 4 so that the new system will have a **unique solution**. Write the new system

```
> eq1 :=
> eq2 :=
> eq3 :=
> graph(eq1,eq2,eq3);
```

Add one equation to those of Activity 4 so that the new system will have **no solution**. Write the new system

```
> eq1 :=
```

```
> eq2 :=
> eq3 :=
> graph(eq1,eq2,eq3);
```

TASK 2

The purpose of this task is to determine algebraically whether a given system of linear equations is consistent or inconsistent. If it is consistent, use backsubstitution to solve the system. The activities use the following system of equations:

$$
\begin{array}{rrrrrrrrrrr}
3x & + & 4y & - & 8z & + & 7t & + & 56m & = & 10 \\
 & & y & + & 9z & - & 7t & + & 100m & = & 9 \\
12x & + & 45y & + & 33z & - & 67t & + & 98m & = & 353 \\
-8x & - & 54y & + & 87z & & & + & 188m & = & 55 \\
x & + & 78y & - & 87z & + & t & & & = & 29
\end{array}
$$

Activity 1

What is the coefficient matrix of the system? (To input a matrix in Maple, type, for example, >A:=matrix([[3,4],[1,2]]);)

```
> A:=
```

Activity 2

What is the augmented matrix of the system?

```
> AUG:=
```

Activity 3

Use the interactive mode of gausselim to obtain the echelon form of AUG (show all the elementary row operations).

Activity 4

Enter the resulting echelon form AUG1 of the original augmented matrix

```
> AUG1 :=
```

By analyzing the matrix AUG1, is the system consistent or inconsistent? If consistent, how many solutions does the system have?

Activity 5

If the system is consistent, apply the interactive mode of backsub to AUG1 to determine the solution.

Activity 6

Change one equation in the system to construct a system that has no solution. Enter the new augmented matrix. Use the nostep mode of `gausselim` function to validate your answer.

Activity 7

What is the minimum number of equations you need to change in the original system so that the new system will have infinitely many solutions? Explain.
     Enter the new system

```
> eq1 :=
> eq2 :=
> eq3 :=
> eq4 :=
> eq5 :=
```

Enter the augmented matrix:

```
> AUG:=
```

Find its echelon form. Again, apply an appropriate algorithm to determine the solution set. Write down the solution. Deduce the solution of the associated homogeneous system.

EXTRA LAB PROBLEMS

1. A certain feed contains 12% protein and 18% carbohydrate. Another feed contains 8% protein and 25% carbohydrate. A mixture containing 10% protein and 20% carbohydrate is needed. How much of each feed should we use to form 300 pounds of the mixture?

2. A firm wants to predict the sale of one of its products. The sales history of the product (in thousands of dollars) in the first three years is as follows:

    | Years | 1 | 2 | 3 |
    | Sale | 10 | 25 | 45 |

    a. One model is to find the equation of the line passing through points (1,10) and (2,25). Find the equation of this line. Does point (3,45) lie on this line? Is it close to the line?

    b. Another model is to find a polynomial of second degree that passes through points (1,10), (2,25) and (3,45). Find this second-degree polynomial.

    c. Which of the two models will give a better prediction for sales in year 4?

Conditions for Consistent Systems

Purpose

The purpose of this lab is to find conditions on parameters which will result in consistent system of linear equations. If consistent, solve the system and interpret the solution.

Automated Linalg Functions

In this lab, you may need to use the following automated functions: `gausselim`, `rref`, `backsub`. For help with any function, type at the Maple prompt: `>?function name`; for example, `>?gausselim`;

Instructions

1. To execute a statement, move the cursor to the line using the mouse or the keyboard and press the enter key.
2. While executing a function, create Maple input regions if needed; otherwise, the output will not appear in the desired place.
3. Execute the following commands to load the package

   ```
   > with(linalg):with(linsys);
   ```

TASK 1

The purpose of this task is to determine conditions on a parameter of a system of linear equations that render a consistent system. The activities use the following system of linear equations:

$$4x + 2y - z = 3$$
$$3x - 5y + 3z = 5/2$$
$$17x + 2y + kz = 8$$

Activity 1

Write the augmented matrix associated with the system.

Activity 2

Apply an appropriate algorithm on the matrix AUG to determine the conditions on the parameter k so that the system has a unique solution or no solution. From the resulting matrix answer the following:

    a. The system has a unique solution if k is equal to _____.

    b. The system has no solution if k is equal to _____.

TASK 2

In modeling some applications, not all constraints of the linear system may be known. The purpose of this task is to determine conditions on parameters of a system of linear equations that render a consistent system. The activities use the following system of linear equations:

$$x + 2y - 11z = d$$
$$2x - 5y + 3z = e$$
$$5x + y - 30z = f$$

Activity 1

Write the augmented matrix AUG associated with the system:

Activity 2

Apply the appropriate algorithm on the matrix AUG to infer conditions on the parameters d, e, and f so that the system is consistent or inconsistent. From the resulting reduced form, respond to the following:

a. The system has infinitely many solutions if d, e and f satisfy the relation, and the solutions are given by _____.

b. The system has no solution if d, e, and f satisfy the relation _____.

Activity 3

Is it possible to change the coefficients x, y, or z so that the resulting system possesses a unique solution? Experiment with several examples and state your conclusion clearly.

EXTRA LAB PROBLEM

Consider the two matrices

```
> A:=matrix([[2,3,4,5],[8,2,5,1],[7,7,5,3]]);
  B:=matrix([[3,6,9,a],[4,5,0,b],[3,1,5,c]]);
```

a. Determine the values of a, b, and c so that matrix A can be transformed into matrix B by a sequence of elementary row operations.

b. Identify the row operations that transfer matrix A to matrix B.

Nutrition Model

Purpose

The purpose of this application is to formulate a mathematical model for a nutrition problem, solve the model, and interpret and analyze the results.

Initialize the package

```
> with(linalg):with(linsys);
```

DESCRIPTION OF THE NUTRITION MODEL

The General Nutrition Center has received an order for 2000 pounds of a food that contains 5% fat, 12% carbohydrates, and 15% protein. They must mix the food using five ingredients. The following table gives the percentages of fat, carbohydrates, protein and the cost per pound for each of the five ingredients a_1, a_2, a_3, a_4, and a_5.

	Fat	Carbohydrates	Protein	Cost/lb($)
a_1	8	5	15	0.55
a_2	6	25	5	0.25
a_3	3	10	20	0.30
a_4	2	15	10	0.35
a_5	4	5	10	0.40

Activity 1

Write the system of linear equations that describes the amount of each ingredient needed to prepare the food.

Activity 2

Enter the augmented matrix AUG of the system of equations of Activity 1.

Activity 3

Is there a unique way of mixing the ingredients to prepare the food? If not, what conditions do you need to impose on each ingredient to have a set of feasible solutions?

Activity 4

Write the cost function associated with the ingredients.

Activity 5

What amounts of each ingredient will give the maximum cost? minimum cost?

Activity 6

The company has an overstock of ingredient a_5 and decides to make up the order using the maximum possible amount of a_5. Find the amounts of each ingredient to prepare this a_5-rich food and the total cost.

Activity 7

If ingredient a_5 is no longer available, what are the amounts needed of the other ingredients to produce the required food?

Oil Refinery Model

Purpose

The purpose of this application is to formulate a mathematical model for an oil refinery problem, solve the model, interpret and analyze the results.

Initialize the package

```
> with(linalg):with(linsys);
```

DESCRIPTION OF THE OIL REFINERY MODEL

An oil company runs four refineries in four different locations. Each refinery produces four products: natural gas(NG); heating oil (HO); diesel oil (DO); and gasoline (G). The number of barrels (in thousands) produced by each refinery and the total allocated cost (in thousands of dollars) of producing the products in each refinery is given in the table.

	NG	DO	HO	G	Cost
R1	30	8	9	8	500
R2	10	35	8	10	800
R3	5	10	35	10	550
R4	10	14	11	35	700

Activity 1

Let x, y, z, w be the fixed costs for producing 1000 barrels of NG, DO, HO, and G, respectively, in each of the refineries. Write the system of linear equations that describes this model.

Activity 2

Write the augmented matrix AUG associated with the system.

Activity 3

Solve for the unknowns x, y, z, and w. Is the solution a feasible one? Explain.

Activity 4

Suppose that the total allocated cost for refinery R1 must be reduced by $50,000. How will this affect the unit cost of each product?

Activity 5

Due to the demand for diesel oil, the company decides that the amount of diesel oil must be twice the amount of heating oil in each of the refineries while keeping allocated costs unchanged in the original table. Will this policy lead to a feasible solution? Explain.

Activity 6

If the answer to Activity 5 is no, the company's policy is to decrease, in the original table, the production of diesel oil by 10% in each of the refineries. Will this change lead to a feasible answer? Explain.

Activity 7

If the answer to Activity 6 is no, the company must modify its policy. The company will notify each refinery to select one product (NG, DO, HO, G) as its major unit of production based on the following criteria. No two refineries can choose the same major product. The number of units each refinery produces must exceed the total combined units of the product produced by the other three refineries. If a refinery already has a major product then it must select that product; otherwise the refinery must increase its selection to meet the criteria.

   Will this policy lead to a solution? If not, experiment with different values of the major product for the refinery that is changing its major production level until a positive solution is attained.

Activity 8

What conditions on the entries of the matrix led to a positive solution? Explain.

Matrix Algebra

Matrices constitute a notational convenience to represent systems of linear equations $Ax = b$ and to tabulate data from experimental results. Thus, it is important to study the properties of matrices, factorization ·algorithms and focus on special matrices that often arise in applications. It is also important to characterize and study properties of matrices that relate to the solvability of systems of linear equations.

LESSON 2.1 Algebra of Matrices

The set of real numbers is closed under the binary operations addition and multiplication; that is, the sum and the product of any two real numbers are also real numbers. Does the set of all $m \times n$ matrices enjoy this property?

Initialize the packages

```
> with(linalg):with(linmat);
```

EQUALITY OF MATRICES

The solution of matrix equations $Ax = b$ requires comparing matrices. Therefore, it is natural to ask: when are two matrices equal?

EXAMPLE 1.1 Consider the two matrices

```
> A:=matrix([[x,x+y],[z,3]]); B:=matrix([[2,6],[2*z,w]]);
```

Are there values of x, y, z, and w for which the two matrices are equal? The two matrices A and B are equal if the following equations are satisfied

```
> eq1:=x=2; eq2:=x+y=6; eq3:=z=2*z; eq4:=3=w;
> solve({eq1,eq2,eq3,eq4},{x,y,z,w});
```

Two matrices A and B are equal if their corresponding entries are equal.

ADDITION OF MATRICES

The set of real numbers is closed under addition; that is, the sum $a + b$ of any two real numbers a and b is a real number. Is the set of all $m \times n$ matrices closed under an appropriate definition of addition?

EXAMPLE 1.2 Consider the two matrices

```
> A:=matrix([[a1,b1],[c1,d1]]);
> B:=matrix([[a2,b2],[c2,d2]]);
```

Their sum is

```
> A+B=evalm(A+B);
```

What is the relation between the entries of the matrix A + B and the corresponding entries of A and B?

EXAMPLE 1.3   Now consider two matrices of larger size

```
> A:=matrix([[a1,b1,c1,d1],[e1,f1,g1,h1]]);
> B:=matrix([[a2,b2,c2,d2],[e2,f2,g2,h2]]);
```

Their sum is

```
> A+B=evalm(A+B);
```

What is the relation between the entries of the matrix A + B and the corresponding entries of A and B?

EXAMPLE 1.4   Consider two matrices of different sizes

```
> A:=matrix([[4,1,2,7],[1,5,3,4]]);
B:=matrix([[1,6,3],[5,6,8]]);
> A+B=evalm(A+B);
```

The error implies that the sum of the two matrices A and B is not well defined. Why?

We can only add matrices of the same size. The resulting matrix is obtained by adding the corresponding entries of the given matrices. The set of all $m \times n$ matrices is closed under addition.

Properties of Matrix Addition

Recall that the set of real numbers under addition satisfies the following:

- **commutative property** $(a + b = b + a)$
- **associative property** $((a + b) + c = a + (b + c))$
- **additive identity property**$(a + 0 = a + 0 = a)$
- **additive inverse property** $(a + (-a) = (-a) + a = 0)$

Does the set of all $m \times n$ matrices enjoy similar properties?

Consider the following set of matrices:

```
> A:=matrix([[a11,a12,a13,a14], a21,a22,a23,a24],
[a31,a32,a33,a34]]);
> B:=matrix([[b11,b12,b13,b14], [b21,b22,b23,b24],
[b31,b32,b33,b34]]);
```

```
> C:=matrix([[c11,c12,c13,c14], [c21,c22,c23,c24],
  [c31,c32,c33,c34]]);
```

EXAMPLE 1.5   Compute the sum A + B

```
> 'A+B'=evalm(A+B);
```

Compare this result to the sum B + A

```
> 'B+A'=evalm(B+A);
```

Are the two matrices A + B and B + A equal?

This example shows that the **commutative property** holds. Can you explain why this statement is true for any two $m \times n$ matrices?

EXAMPLE 1.6   Compute the sum (A + B) + C

```
> '(A+B)+C'=evalm(evalm(A+B)+C);
```

Compare this result with the sum A + (B + C)

```
> 'A+(B+C)'=evalm(A+evalm(B+C));
```

Are the two matrices (A + B) + C and A + (B + C) equal?

This example shows that the **associative property** holds. Can you explain why this statement is true for any three $m \times n$ matrices?

EXAMPLE 1.7   Is there a matrix Z such that A + Z = A for any matrix A? (*Hint*: a + 0 = a for any real number a.) The matrix Z is the **Zero matrix** all whose entries are zero.

EXAMPLE 1.8   Is there a matrix X such that A + X = Z, where Z is the zero matrix? Try the following

```
> '-1*A'=evalm( (-1)*A);
> Z:=evalm(evalm(A)+evalm((-1)*A));
```

(−1) ∗ A (or −A) is called the **additive inverse** of A. How do you construct the additive inverse of a given matrix?

SUMMARY
The set of all $m \times n$ matrices is closed under addition.
The commutative property holds:
$$A + B = B + A$$
The associative property holds:
$$(A + B) + C = A + (B + C)$$
The zero matrix Z is the identity element under addition:
$$A + Z = Z + A = A$$

The additive inverse of the matrix A is -A:
$$A + (-A) = (-A) + A = Z$$

MULTIPLICATION BY A SCALAR

What is the effect of multiplying a matrix by a real scalar?

EXAMPLE 1.9 Consider the matrix

```
> A:=matrix([[1,2],[3,4]]);
```

Multiply matrix A by the scalar k

```
> k*A=evalm(k*A);
```

EXAMPLE 1.10 Consider the matrix

```
> A:=matrix([[1,2,3],[3,4,5],[5,7,9]]);
> k*A=evalm(k*A);
```

How do entries of the matrix kA relate to the entries of the matrix A?

The result of multiplying a matrix by a scalar is a matrix whose entries are obtained by multiplying each entry of the given matrix by the given scalar.

Properties of Multiplication by a Scalar

EXAMPLE 1.11 Consider the following matrices:

```
> A:=matrix([[a11,a12,a13,a14],[a21,a22,a23,a24],
[a31,a32,a33,a34]]);
> B:=matrix([[b11,b12,b13,b14],[b21,b22,b23,b24],
[b31,b32,b33,b34]]);
```

Compute the scalar multiples k_1A and k_2A and their sum

```
> k1A:=evalm(k1*A); k2A:=evalm(k2*A);
> k1A+k2A=evalm(k1A+k2A);
```

Compare the sum $k_1A + k_2A$ to $(k_1 + k_2)A$

```
> (k1+k2)*A=evalm((k1+k2)*A);
```

Is $k_1A + k_2A$ equal to $(k_1 + k_2)A$ for any matrix A? Can you explain why this equality holds for any $m \times n$ matrix A and any scalars k_1 and k_2?

Is $(kA + kB)$ equal to $k(A + B)$ for any matrices A and B?

EXAMPLE 1.12   Consider the following matrices:

```
> A:=matrix([[a11,a12,a13,a14],[a21,a22,a23,a24],
[a31,a32,a33,a34]]);
> B:=matrix([[b11,b12,b13,b14],[b21,b22,b23,b24],
[b31,b32,b33,b34]]);
```

Compute the scalar product k(A+B)

```
> k(A+B) := evalm(k*(evalm(A+B)));
```

Compare the result to the sum kA + kB

```
> kA+kB =evalm(evalm(k*A)+evalm(k*B));
```

Can you explain why the equality holds for any $m \times n$ matrix A and any scalar k?

Is $(k_1 k_2)A$ equal to $k_1(k_2 A)$ for any matrix?
Can you explain why this equality holds for any $m \times n$ matrix A and any scalars k_1 and k_2?

MATRIX MULTIPLICATION

The set of real numbers is closed under multiplication; that is, the product $a * b$ is a real number for any two real numbers a and b. Is this the case for the set of all $m \times n$ matrices?

EXAMPLE 1.13   Consider the matrices

```
> A:=matrix([[4,1,2,7]]); B:=matrix([[1],[6],[3]]);
```

A is a 1×4 matrix and B is 3×1 matrix. Consider the product BA

```
> BA=multiply(B,A);
```

What does this tell us about product BA? Product BA is a 3×4 matrix obtained by multiplying a 3×1 matrix B and a 1×4 matrix A. What if we multiply A and B?

```
> AB=multiply(A,B);
```

In this case product AB is not defined.

EXAMPLE 1.14   Consider now the matrices

```
> A:=matrix([[a1,a2,a3]]); B:=matrix([[b1],[b2],[b3]]);
> AB=multiply(A,B);
```

The result is a real number.

Is matrix multiplication always well-defined? Is the product of two matrices (whenever defined) a matrix? Is product AB equal to product BA for any two matrices A and B?

How does this multiplication process work? Use the function `matrixmul` to observe the steps.

EXAMPLE 1.15 Consider the two matrices

```
> A:=matrix([[a11,a12,a13],[a21,a22,a23],
[a31,a32,a33]]);
> B:=matrix([[b11,b12],[b21,b22],[b31,b32]]);
```

The product AB is

```
> matrixmul(A,B);
```

Compare this with the Maple output

```
> AB = multiply(A,B);
```

EXAMPLE 1.16 Consider the two matrices

```
> A:=matrix([[1,2,3],[2,2,3],[3,-1,0]]);
> B:=matrix([[1,2],[2,2],[3,2]]);
```

The product AB is

```
> matrixmul(A,B);
```

Compare this with the Maple output

```
> AB=multiply(A,B);
```

The product of an $m \times n$ matrix A with an $n \times r$ matrix B is obtained by multiplying row i of A with column j of B and substituting the result in (i,j)– entry of the product matrix for each i $(i = 1 \ldots m)$ and j $(j = 1 \ldots r)$. The result is an $m \times r$ matrix.

Application of Matrix Multiplication

Matrix multiplication may be viewed as a transformation of a given image from one position into another. For example, suppose we are given the two dimensional image defined by the following set of points:

```
> S:={[1,2],[1.2,2.1],[1.22,2.51],[1.4,3.2],
[1.45,3.11],[1.5,2.3]};
```

Call the function `Geometry` that will transform the image S to another image S_1 based on multiplication by the matrix:

```
> T:=matrix([[1,0],[0,-1]]);
> Geometry(T,S);
```

As another example, consider the two dimensional image of a semicircle generated by the code

```
> S:={}:n:=40: for i to n do t:=(i-1)/n:
S:=S union {[t,sqrt(1-t^2)]}:od:
```

Call the function `Geometry` that will transform the image S to another image S_1 based on the multiplication by the matrix:

```
> T:=matrix([[-1,0],[0,1]]);
> Geometry(T,S);
```

Properties of Matrix Multiplication

The set of all nonzero real numbers under multiplication satisfies the **commutative property** (ab = ba), the **associative property** ((ab)c = a(bc)), the **distributive property** (a(b + c) = ab + bc), the **multiplicative identity property** (a.1 = 1.a = a), and the **multiplicative inverse property** (a.$\frac{1}{a}$ = 1 provided a ≠ 0). Does the set of all $m \times n$ matrices enjoy similar properties?

EXAMPLE 1.17   Consider the two matrices

```
> A:=matrix([[a11,a12],[a21,a22]]);
B:=matrix([[b11,b12],[b21,b22]]);
```

Is AB = BA?

```
> AB = multiply(A,B);
> BA = multiply(B,A);
```

In general, AB is not equal to BA.

Check whether the following properties hold by choosing three matrices A, B, and C:

- **associative property** (AB)C = A(BC)
- **distributive property** A(B + C) = AB + AC

SUMMARY
Matrix multiplication is not always defined.
The commutative property does not always hold.
The associative property holds:
(AB)C = A(BC).
The distributive property holds:
A(B + C) = AB + AC.

We still have to elaborate on the **multiplicative inverse property** for matrices. For a scalar equation ax = b in one unknown x, the solution is x = $\frac{b}{a}$ or x = $a^{-1}b$, provided a is not equal to zero. What is an equivalent statement for a matrix equation Ax = B?

EXERCISES

In these exercises you may need to use the following Maple commands

- To **enter a matrix**

  ```
  > S:={[a,b,c],[1,2,3],[4,6,7]};
  ```

- To **add two matrices**

  ```
  > C:=evalm(A+B);
  ```

- to **multiply two matrices**

  ```
  > C:=multiply(A,B);
  ```

- To **multiply a matrix A by a scalar k**

  ```
  > C:=evalm(k*A);
  ```

- To **extract a row (column) of a matrix** A

  ```
  > r[i]:=row(A,i); c[i]:=col(A,i);
  ```

1. Consider the matrix

   ```
   > A:=matrix([[0,1],[-1,-1]]);
   ```

   Give a general formula for computing powers of matrix A for any positive integer n. *Hint*: Compute a few powers of A and deduce the formula.

2. Consider the matrix

   ```
   > A:=matrix([[x,1],[0,x]]);
   ```

   By computing several powers of A, deduce a general formula for the nth power, A^n, of A. Can you describe what happens to the entries of A^n when $n \to \infty$?

3. Consider the matrix

   ```
   > A:=matrix([[0.35,0.35,0.35], [0.25,0.25,0.25],
       [0.4,0.4,0.4]]);
   ```

   By computing a few powers of matrix A, deduce the value of the nth power, A^n, as $n \to \infty$. Can you construct a matrix B whose entries behaves like matrix A? Explain the construction process.

4. Consider the matrices

   ```
   > A:=matrix([[1,2],[3,-1]]); I2:=diag(1,1);
   ```

   Find the coefficients a, b, and c of the polynomial p(x)

   ```
   > p:=x->a*x^2+b*x+c;
   ```

such that $p(A) = 0$ where $p(A) = aA^2 + bA + cI$. Is this polynomial unique?

5. Consider the matrix

```
>   A:=matrix([[1,0],[0,-1]]);
```

The point (1,2) in the x–y plane can be represented as 2×1 matrix

```
> P:=matrix([[1],[2]]);
```

Find product AP

```
> multiply(A,P);
```

Determine the position of the resulting point. How does it compare with the original point (1,2)?

Special Types of Matrices

In this lesson we introduce the properties of special types of matrices that will be visited throughout the rest of the course.

Initialize the packages

```
> with(linalg):with(linmat);
```

DIAGONAL MATRIX

A diagonal matrix is a square matrix whose off diagonal entries are all zero. The Maple command that generates, for example, a 3×3 diagonal matrix with diagonal elements 1,2, and 5 is

```
> D1:=diag(1,2,5);
```

The command for a 4×4 diagonal matrix with entries a, b, c, and d is

```
> D2:=diag(a,b,c,d);
```

IDENTITY MATRIX

An identity matrix is a diagonal matrix with diagonal entries all equal to 1. The 3×3 identity matrix is

```
> I3:=diag(1,1,1);
```

and the 4×4 identity matrix is

```
> I4:=diag(1,1,1,1);
```

Recall that in the set of real numbers, the number 1 plays the role of the **multiplicative identity**; that is, for any real number a, $a * 1 = 1 * a = a$. Is there an analogous statement for matrices? What is the result of multiplying any square matrix A with its associated identity matrix?

EXAMPLE 2.1   Consider a 3×3 matrix A and the 3×3 identity matrix

```
> A:=matrix([[a,b,c],[d,e,f],[g,h,i]]); I3:=diag(1,1,1);
> multiply(A,I3);
```

The identity matrix plays the role of the multiplicative identity.

Before we proceed to discuss other types of matrices, we introduce the **transpose of a matrix.**

The transpose of an $m \times n$ matrix A is obtained by interchanging the rows and columns of A.

EXAMPLE 2.2   Consider the matrix

```
> A:=matrix([[1,2,3],[2,2,4],[0,1,5]]);
```

The transpose of matrix A is

```
> B:=transpose(A);
```

The transpose of the transpose of A is

```
> C:=transpose(B);
```

Consider a 3×4 matrix A with general entries

```
> A:=matrix([[a11,a12,a13,a14],[a21,a22,a23,a24],
[a31,a32,a33,a34]]);
> A^T=transpose(A);
> (A^T)^T=transpose(transpose(A));
```

What is the relation between A and $(A^T)^T$? Check using the following function

```
> transtrans(A);
```

EXAMPLE 2.3   Consider the two matrices

```
> A:=matrix([[a11,a12,a13],[a21,a22,a23],[a31,a32,a33]]);
B:=matrix([[b11,b12,b13],[b21,b22,b23],[b31,b32,b33]]);
```

Is $(A + B)^T = A^T + B^T$? Check using the following function

```
> trsum(A,B);
```

What do you conclude? Is this statement true in general?

Is $(AB)^T = B^T A^T$ for the following 2×2 matrices A and B?

```
> A:=matrix([[a11,a12],[a21,a22]]);
B:=matrix([[b11,b12],[b21,b22]]);
```

Check using the following function

```
> trproduct(A,B);
```

What do you conclude? Is this statement true in general?

SYMMETRIC MATRICES

Symmetric matrices arise in the formulation of many physical and engineering problems. Let us look at the class of all matrices A that satisfy the property $A = A^T$.

EXAMPLE 2.4   Consider the matrix

```
> A:=matrix([[1,2,5],[2,6,9],[5,9,11]]);
```

The transpose of A is

```
> A^T=transpose(A);
```

In general, what relations must be satisfied by the entries of a matrix A that will make $A = A^T$?

Come up with more examples to formulate a general statement about symmetric matrices, matrices that are equal to their transpose.

TRIANGULAR MATRICES

Triangular matrices arise naturally in the formulation of many modeling problems. These matrices are called upper or lower depending on whether the entries below or above the main diagonal are all zero. An **upper triangular matrix** is

```
> A :=matrix([[1,2,4],[0,5,0],[0,0,8]]);
```

A **lower triangular matrix** is

```
> B:= matrix([[1,0,0],[2,5,0],[0,6,0]]);
```

The row echelon form of a matrix is another example of an upper triangular matrix. Consider the matrix

```
> A:=matrix([[2,3,5,6],[4,7,8,9],[9,7,5,3],[1,3,6,9]]);
```

Obtain its row echelon form

```
> B:=gausselim(A);
```

Is the product of two upper triangular matrices an upper triangular matrix?

EXAMPLE 2.5   Consider the two matrices

```
> A:=matrix([[1,3,5,1],[0,3,4,5],[0,0,1,4],[0,0,0,7]]);
> B:=matrix([[3,1,0,1],[0,9,1,1],[0,0,1,3],[0,0,0,3]]);
```

The product is

```
> multiply(A,B);
```

Is the transpose of a triangular matrix a triangular matrix?

EXAMPLE 2.6   Consider the matrix

```
> A:=matrix([[1,3,5,1],[0,3,4,5],[0,0,1,4],[0,0,0,7]]);
```

The transpose of A is

```
> B:=transpose(A);
```

What about the sum of two upper (lower) triangular matrices? Is it upper (lower) triangular?

ELEMENTARY MATRICES

Matrices that are obtained by applying a *single* elementary row operation to the identity matrix are called elementary matrices.

EXAMPLE 2.7   Consider the 3×3 identity matrix

```
> I3:=diag(1,1,1);
```

The matrix

```
> E1:=swaprow(I3,1,2);
```

is obtained from the identity matrix I_3 by interchanging the first and second row. The matrix

```
> E2:=mulrow(I3,3,2);
```

is obtained from the identity matrix I_3 by multiplying the third row by the scalar 2. The matrix

```
> E3:=addrow(I3,2,3,4);
```

is obtained from the identity matrix I_3 by adding four times the second row to the third row. Each of the matrices E_1, E_2, and E_3 is obtained by a *single* application of an elementary row operation.

Are the following matrices examples of elementary matrices?

```
> E4:=diag(1,3,4);
> E5:=matrix([[1,0,0,0],[0,1,0,0],[0,1,1,0],[0,0,0,1]]);
```

Matrix E_4 is not an elementary matrix because E_4 is obtained by applying two elementary row operations to the identity matrix. E_5 is an elementary matrix because it was obtained from the identity by applying a single elementary row operation.

EXERCISES

In the following exercises you may need to use the automated functions `trproduct`, `transtrans`, `trsum`, `transpose`, and the Maple commands `mulrow(A,i,j,n)`, `swaprow(A,i,j)`, and `gausselim`.

1. Consider a 3×3 matrix A with random entries obtained by employing the Maple command

```
> A:=randmatrix(3,3);
```

Compute the product of A and its transpose. Is the resulting matrix a symmetric matrix? If so, can you give an argument to support your assertion?

2. Consider matrix A whose (i,j)-th entry is given by $A[i,j] = \frac{1}{i+j-1}$. Such a matrix is called a **Hilbert matrix**. A 3×3 Hilbert matrix can be generated using the following Maple procedure:

```
> A:=matrix(3,3):
> for i from 1 to 3 do for j from 1 to 3 do
    A[i,j]:=1/(i+j+1); od;od; print('A=', A);
```

    a. Is the Hilbert matrix a symmetric matrix?
    b. Solve the system $Ax = b$, where

```
> b:=matrix([[0],[1],[2]]);
```

Write the augmented matrix, obtain the echelon form, and apply backsubstitution.

3. Which of the following are examples of elementary matrices and why?

```
> A1:=matrix([[1,0,2],[0,1,0],[0,0,1]]);
> A2:=matrix([[1,0,0],[0,7,0],[0,0,1]]);
> A3:=matrix([[1,0,0],[1,1,0],[3,0,1]]);
> A4:=matrix([[1,0,0],[0,1,-4],[0,0,1]]);
```

4. Consider the matrix

```
> A:=matrix([[1,2,3,4],[4,1,2,3],[3,4,1,2],[2,3,4,1]]);
```

Multiply matrix A with matrix v:

```
> v:=matrix([[1],[1],[1],[1]]);
```

How do the entries of the resulting matrix relate to the rows of A?
Experiment with matrices of different sizes to propose a general statement regarding this relation. Such matrices are called **circulant matrices**.

5. Give an example to show that the elementary row operations applied to a matrix A are equivalent to premultiplying (that is multiplying the matrix from left) matrix A by elementary matrices obtained by similar row operations.

Singular Matrices

Singular matrices arise in the formulation of many engineering problems. Properties of singular matrices shed light on the solvability of systems of linear equations.

Initialize the packages

```
> with(linalg):with(linmat);
```

COEFFICIENT MATRIX OF NONHOMOGENEOUS SYSTEMS

Let us start with examples of matrices associated with systems of linear equations with an equal number of equations as unknowns.

EXAMPLE 3.1   The augmented matrix of a nonhomogeneous system of linear equations is

```
> AUG:=matrix([[1,3,6,7],[1,5,8,4],[3,13,22,15]]);
```

Apply Gauss elimination to AUG

```
> B:=gausselim(AUG);
```

What does matrix B tell us about the solution set of the system of equations represented by AUG?

Let us apply backsubstitution to solve the system

```
> backsub(B);
```

The system is consistent with infinitely many solutions. For this system, let us investigate the associated coefficient matrix

```
> A:=matrix([[1,3,6],[1,5,8],[3,13,22]]);
```

Apply Gauss elimination to A

```
> B1:=gausselim(A);
```

This example shows that the underlying system has **infinitely many solutions**. Note that the matrix B_1, in echelon form corresponding to the coefficient matrix A, has a zero row.

EXAMPLE 3.2   Alter the matrix AUG slightly. Consider the augmented matrix

```
> AUG:=matrix([[1,3,6,7],[1,5,8,4],[3,13,22,1]]);
```

Apply Gauss elimination to AUG

```
> B:=gausselim(AUG);
```

What does matrix B tell us about the solution set of the system of equations represented by AUG?
Let us apply backsubstitution to solve the system

```
> backsub(B);
```

For this system, let us investigate the associated coefficient matrix. The coefficient matrix A is

```
> A:=matrix([[1,3,6],[1,5,8],[3,13,22]]);
```

Apply Gauss elimination

```
> B1:=gausselim(A);
```

The slight modification in the augmented matrix leads to a system that is **inconsistent**. Note that the matrix B_1, in echelon form corresponding to the coefficient matrix A, has a zero row.

EXAMPLE 3.3   Now alter the matrix AUG of Example 3.2. Consider the augmented matrix

```
> AUG:=matrix([[1,3,6,7],[1,5,8,4],[3,13,2,1]]);
```

Apply Gauss elimination to AUG

```
> B:=gausselim(AUG);
```

What does matrix B tell us about the solution set of the system of equations represented by AUG?
Let us apply backsubstitution to solve the system

```
> backsub(B);
```

For this system, let us investigate the associated coefficient matrix

```
> A:=matrix([[1,3,6],[1,5,8],[3,13,2]]);
```

Apply Gauss elimination

```
> B1:=gausselim(A);
```

This slight modification leads to a system that has a **unique solution**. Note that the last row of matrix B_1 in echelon form corresponding to the coefficient matrix A, is not zero.

In Examples 3.1–3.3, the coefficient matrix associated with the augmented matrix AUG determines whether the system represented by AUG has a unique solution or not.

Recall that in solving the scalar equation $a * x = b$, where a and b are real numbers, one of the following rules must hold:

- If $a = 0$ and $b = 0$, the equation has infinitely many solutions.
- If $a = 0$ and b is not equal to zero, the equation has no solution.
- If a is not equal to zero, the equation will have a unique solution given by $x = a^{-1} * b$ (where a^{-1} is the multiplicative inverse of a).

For systems of linear equations $AX = B$, where A and B are known matrices and X is the unknown quantity, we have the following analogous situation:

As in Example 3.1, the system of linear equations has infinitely many solutions, and in Example 3.2 the system has no solution. In these two examples, we say that the coefficient matrix is **singular**. Note that the reduced matrix B_1 in echelon form has at least one zero row.

In Example 3.3, the system of linear equations has a unique solution and the coefficient matrix is **nonsingular**. Analogous to the scalar case, matrix A must have a multiplicative inverse. Note that the reduced matrix B_1 in echelon form has no zero row.

NONSINGULAR MATRIX

EXAMPLE 3.4   Consider the matrix

```
> A:=matrix([[2,4,6],[4,6,7],[8,5,2]]);
```

The inverse of A is

```
> B:=linalg[inverse](A);
```

To verify this,

```
> AB=multiply(A,B);
> BA=multiply(B,A);
```

An $n \times n$ matrix A is a nonsingular matrix if and only if there is an $n \times n$ matrix B so that the product $AB = BA = I$ where I is the identity matrix. Otherwise, A is called a singular matrix. Matrix B is called the inverse of A.

If an $n \times n$ matrix has an inverse, then the inverse of the matrix is unique. (Fact 2.1)

EXAMPLE 3.5   Consider the matrix

```
> A:=matrix([[1,2,0],[0,1,1],[1,2,1]]);
```

Its inverse A_1 is

```
> A1:=linalg[inverse](A);
```

Is $A = (A^{-1})^{-1}$?

```
> inv(inv(A)) =linalg[inverse](A1);
```

If A is an $n \times n$ invertible matrix, then $(A^{-1})^{-1} = A$. (Fact 2.2)

Consider another nonsingular matrix B and its product with matrix A.

```
> B:=matrix([[0,2,1],[1,1,0],[1,2,1]]);
> C:=multipy(A,B);
```

Take the inverse of matrices A and B. Let $A_1 = A^{-1}$ and $B_1 = B^{-1}$.

```
> A1:=linalg[inverse](A);
> B1:=linalg[inverse](B);
```

What is the relation between the inverse of the product AB and the product of the inverses of A and B?

Is the product of two nonsingular matrices nonsingular?

```
> inv(A)*inv(B)=multiply(A1,B1);
> inv(B)*inv(A)=multiply(B1,A1);
> inv(AB)=evalm(linalg[inverse](C));
```

Compare $A^{-1}B^{-1}$ and $B^{-1}A^{-1}$ to the inverse of the product AB. What do you conclude?

If A and B are two nonsingular matrices, then the inverse of the product AB is equal to the inverse of B times the inverse of A; that is, $(AB)^{-1} = B^{-1}A^{-1}$. (Fact 2.3)

EXAMPLE 3.6   Take the following two matrices

```
> A:=matrix([[5,6,7],[5,7,2],[6,7,11]]);
> B:=matrix([[5,6,7],[5,7,2],[15,19,16]]);
```

and find a condition under which the multiplicative inverse exists. Obtain the reduced echelon forms of the matrices A and B

```
> A1:=rref(A);
> B1:=rref(B);
```

Is there a difference between the reduced row echelon form of matrix A and the reduced echelon form of matrix B?

Matrix A is row equivalent to the identity matrix. That is, by applying a sequence of elementary row operations to matrix A, we obtain the identity matrix. Matrix B is not row equivalent to the identity matrix. Which one of them has an inverse?

```
> linalg[inverse](A);
> linalg[inverse](B);
```

Matrix A has an inverse while the inverse of the matrix B does not exist.

An $n \times n$ matrix A is invertible if and only if it is row equivalent to the identity matrix I_n (Fact 2.4)

NONSINGULAR MATRICES AND SYSTEMS OF LINEAR EQUATIONS

EXAMPLE 3.7   Consider the homogeneous system of linear equations whose augmented matrix is

```
> HS:=matrix([[5,6,7,0],[5,7,2,0],[6,7,11,0]]);
```

The solution can be obtained by performing Gauss elimination and then applying the backsubstitution algorithm

```
> HS1:=gausselim(HS);
> backsub(HS1);
```

The homogeneous system of equations has the **trivial solution**. What can you say about the coefficient matrix A of HS? For example, does the inverse of A exist?

```
> A:=matrix([[5,6,7],[5,7,2],[6,7,11]]);
```

Its inverse is

```
> linalg[inverse](A);
```

EXAMPLE 3.8   Consider a nonhomogeneous system NHS associated with the invertible matrix A of Example 3.7. The augmented matrix is

```
> NHS:=matrix([[5,6,7,1],[5,7,2,15],[6,7,11,23]]);
```

The solution can be obtained by performing Gauss elimination and then applying the backsubstitution algorithm.

```
> NHS1:=gausselim(NHS);
> backsub(NHS1);
```

This system has a **unique solution**.

If the coefficient matrix is nonsingular (invertible), then the nonhomogeneous system of linear equations has a unique solution. (Fact 2.5)

If the coefficient matrix is nonsingular (invertible), then the only solution of the homogeneous system of linear equations is the trivial solution. (Fact 2.6)

SUMMARY

An $n \times n$ matrix A is nonsingular (invertible)
if and only if
A is row equivalent to the identity matrix
if and only if

the homogeneous system whose coefficient matrix is A has only the zero solution

if and only if

the nonhomogeneous system whose coefficient matrix is A has a unique solution

---

EXERCISES

In the following exercises you may need to use the automated functions `gausselim`, `backsub`, `linalg[inverse](A)`, and `multiply(A,b)`.

1. Consider the following system of linear equations:

```
> eq1:=x-2*y+3*z-w=1;
> eq2:=2*x-y+z-3*w=1/2;
> eq3:=3/2*x-5*y-6*z=2;
> eq4:=x-4*y+6*z-7*w=-3;
```

  a. Write the coefficient matrix A of the system.
  b. Is matrix A nonsingular?
  c. If A has an inverse, use the proper Maple command to find the inverse.
  d. Use the information in part (c) to deduce the solution of the system
  e. Use `gausselim` and `backsub` to solve the system. Confirm your answer to part (d).

2. Consider the following system of linear equations:

```
> eq1:=x-2*y+3*z=1;
> eq2:=2*x-y+3*z=-5;
> eq3:=x-5*y+c*z=2;
```

  a. Write the augmented matrix A of the system.
  b. Obtain the echelon form of the matrix A.
  c. Choose a value for c for which the system has (i) no solution and (ii) has a unique solution.
  d. For each value of c that you chose, write the corresponding matrix.
  e. Which one of these matrices is nonsingular?
  f. State in your own words the relations between the results of (c) and (e).

3. Obtain the reduced echelon form of the following matrices to determine whether their inverses exist:

```
> A1:=matrix([[-1,2,-3,5],[7,9,1,2],[5,-9,-5,3],[8,5,6,4]]);
> A2:=matrix([[-1,2,-3,5,6],[7,9,1,2,11],[5,-9,-5,3,12],
      [11,2,-7,10,29],[-1,2,4,6,9]]);
```

State the reasons for your conclusions.

4. Find the matrix A whose inverse is given by the matrix

```
> B:=matrix([[1,-1,0,0,0],[0,1,-1,0,0],[0,0,1,-1,0],
    [0,0,0,1,-1],[0,0,0,0,1]]);
```

5. By choosing your own matrices and running a few experiments, decide which of the following statements are incorrect. Justify your answer. Based on your observations, can you state a correct version of the statement?

   a. Any elementary matrix is invertible.
   b. The inverse of a symmetric matrix is symmetric.
   c. The inverse of an upper triangular matrix is a lower triangular matrix.
   d. The inverse of a lower triangular matrix is an upper triangular matrix.
   e. The inverse of an elementary matrix is an elementary matrix.

6. Give examples of matrices A and B to determine a relation between $(AB)^{-1}$ and the product of A^{-1} and B^{-1}.

   a. Enter the two matrices A and B.
   b. Compute the relevant quantities to deduce your answer.
   c. What is the relation between $(AB)^{-1}$ and the product of A^{-1} and B^{-1}?

   Can you verify your conclusion?

7. Give an example of a matrix A to determine a relation between the inverse of the transpose of A and the transpose of the inverse of A.

   a. Enter the matrix

   ```
   > A:=
   ```

   b. Compute the relevant quantities to deduce your answer.
   c. What is the relation between the $(A^{-1})^T = (A^T)^{-1}$?
   d. Can you verify your conclusion?

8. Investigate the properties of upper and lower triangular matrices.

   a. Is the product of two upper triangular matrices an upper triangular? Enter two 3×3 upper triangular matrices. Compute the product. What is your conclusion?
   b. Is the product of two lower triangular matrices a lower triangular? Enter two 3×3 lower triangular matrices. Compute the product. What is your conclusion?
   c. Is the product of a lower triangular and an upper triangular matrix a lower triangular or an upper triangular? Enter two 3×3 matrices, one upper triangular and one lower triangular. Compute their product. What is your conclusion?

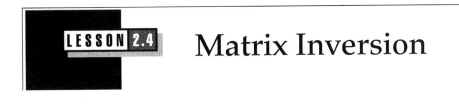

Matrix Inversion

One algorithm often used to construct the inverse of a matrix is the Gauss-Jordan elimination procedure. But before we construct the inverse, we need to check whether the inverse of the matrix exists. Based on our discussion in Lesson 2.3, an $n \times n$ matrix A is invertible if and only if it is row equivalent to the identity.

Initialize the packages

```
> with(linalg): with(linmat);
```

FINDING THE INVERSE USING GAUSS-JORDAN ELIMINATION

The steps in the Gauss-Jordan elimination procedure are

1. Augment the given $n \times n$ matrix A with the identity matrix I_n of the same dimension; that is, form the matrix $[A : I_n]$.

2. Apply a sequence of elementary row operations to convert matrix A to the identity matrix. The same sequence of operations convert the identity matrix to the desired inverse of the matrix A.

Learning the Process

Apply the demonstration mode of the function `inverse` to go through the steps of finding the inverse of A, choosing the Gauss-Jordan procedure from the menu.

EXAMPLE 4.1 Consider the matrix

```
> A:=matrix([[1,4,6],[6,4,9],[1,7,9]]);
> inverse(A);
```

Repeat using as many of your own matrices as necessary to master the process.

INVERSES AND SYSTEMS OF LINEAR EQUATIONS

As noted in Example 3.8, if the coefficient matrix of a system of linear equations $Ax = b$ is invertible, then the system has a unique solution given by $A^{-1}b$.

EXAMPLE 4.2   Consider the system of equations

```
> eq1:=x1+4*x2+6*x3=11;eq2:=6*x1+4*x2+9*x3=19;
eq3:=x1+7*x2+9*x3=35;
```

This system can be written $Ax = b$ where

```
> A:=matrix([[1,4,6],[6,4,9],[1,7,9]]);
> b:=matrix([[11],[19],[35]]);
```

Check whether the inverse of matrix A exists (use the nostep mode)

```
> inverse(A);
```

Thus, the inverse of A is

```
> B:=matrix([[-9/7,2/7,4/7],[-15/7,1/7,9/7],
[38/21,-1/7,-20/21]]);
```

Since the inverse of matrix A is matrix B, multiply both sides of $Ax = b$ by B to get the solution

```
> x:=multiply(B,b);
```

This solution coincides with the solution that we get directly from

```
> solve({eq1,eq2,eq3},{x1,x2,x3});
```

APPLICATION IN CRYPTOGRAPHY

An application of matrix inversion arises in cryptography, the study of encoding and decoding of messages.

EXAMPLE 4.3   Suppose we want to send the message "LINEAR ALGEBRA TOOL." We assume that the letters $A, B, C, \ldots, Z$ are mapped into the numbers $1, 2, 3, \ldots, 26$. For example, the number 12 represents the letter L. A period is mapped into a number greater than 26, say the number 30. We assume that the messages contain only numbers. Therefore, the set of numbers

$$\{12, 9, 14, 5, 1, 18, 30, 1, 12, 7, 5, 2, 18, 1, 30, 20, 15, 15, 12, 30\}$$

represents the message "LINEAR.ALGEBRA.TOOL." This message in matrix form is

```
> M:=matrix([[12,1,12,18,15],[9,18,7,1,15],
[14,30,5,30,12],[5,1,2,20,30]]);
```

The encoding matrix, known to the sender and to the receiver, is

```
> C:=matrix([[-1,1,2,1],[0,2,-1,1],
  [0,0,4,1],[1,-3,2,0]]);
```

The encoded message goes through a channel whose output is the product CM. In this setting, the receiver will get the message represented by the matrix

```
> R:=multiply(C,M);
```

How does the receiver decode the message on receiving matrix R? To decode the message, the receiver multiplies the encoded message R, by the inverse of the encoding matrix C. The inverse of the encoding matrix C is

```
> C1:=linalg[inverse](C);
```

The receiver decodes the message by computing the product of C_1 and R

```
> multiply(C1,R);
```

Successively converting the entries of this matrix space columnwise reproduces the original message.

EXAMPLE 4.4   Decode the following message

$$R := \{36, -10, 88, 50, 30, 45, 24, -32, 8, 17, 16, 2, 34, 37, 110, 44, 46, 69, 70, 0\}$$

using the encoding matrix C of Example 4.3. The matrix representing the message given by the set R is

```
> R:=matrix([[36,30,8,34,46],[-10,45,17,37,69],
  [88,24,16,110,70],[50,-32,2,44,0]]);
```

To decode the message, multiply the matrices C^{-1} and R. First find the inverse of C

```
> C1:=linalg[inverse](C);
```

and the product $C^{-1}R$

```
> multiply(C1,R);
```

to decode the message. What was the message?

EXERCISES

In the following exercises you may need to use the automated functions `inverse` and the Maple commands `gausselim`, `backsub`, `rref`, and `solve`

1. The diagonal matrix

```
> A:=diag(1,2,3,4,5,6);
```

Find the inverse of A. Can you come up with a general statement for finding the inverse of any $n \times n$ diagonal matrix? Can you verify your statement?

2. The matrix

```
> A:=matrix([[-7/5,-6/5,4/5,8/5],[-6/5,-3/5,2/5,4/5],
    [4/5,2/5,-3/5,-6/5],[8/5,4/5,-6/5,-7/5]]);
```

a. Find the inverse A_1 of A.
b. Compute the inverse of the first 10 powers of the matrix A.
c. Compute the first 10 powers of the matrix A.
d. Compare the results of parts (b) and (c). Can you deduce a general relation? If so, verify your answer.

3. Consider a matrix generated via the simple Maple procedure

```
> A:=matrix(5,5):
> for i from 1 to 5 do for j from 1 to 5 do
    A[i,j]:=1/(i+j-1);od;od; print('A=',A);
```

a. Find A^{-1}.
b. Show that the sum of the entries of the ith row and ith column is the same for A^{-1}.

4. Consider the matrices

```
> A:=matrix([[2,-1,0,0],[-1,2,-1,0],[0,-1,2,-1],[0,0,-1,2]]);
> B:=diag(3,3,3,3);
```

a. Compute $A + B$. Find the inverse $(A + B)^{-1}$.
b. Compute the product $A(A + B)^{-1}B$.
c. Compute the inverses A^{-1} and B^{-1}.
d. Compute the sum $A^{-1} + B^{-1}$.
e. Compute the inverse of the sum in part (d).
f. Compare the results of parts (b) and (e).
g. Deduce a formula. Verify this formula.

5. Use the encoding matrix

```
> C:=matrix([[-1,1,2,1,1,1],[0,1,0,2,-1,1],
    [1,-1,0,0,4,1],[1,-3,0,1,2,0],[2,-1,2,3,-1,2],
    [0,1,-4,5,6,3]]);
```

to decode the encrypted message

$$\{49, 36, 25, -8, 93, 38, -1, 12, 61, 54, 123, 24, 131, 40, 133, 62,$$

$$203, 120, 59, 12, 55, 24, 77, 36, 98, 81, 49, -31, 115, 216\}$$

What is the message?

Determinant of a Matrix

The notion of determinants is related to the solution of linear systems. Indeed, the determinant of a matrix is a measure that may be used to determine the consistency of the square linear system associated with the matrix.

Initialize the packages

```
> with(linalg):with(linmat);
```

DETERMINANTS

EXAMPLE 5.1   Consider the system of linear equations

```
> eq1:= a*x+b*y=f1;eq2:=c*x+d*y=f2;
```

On solving the system

```
> solve({eq1,eq2},{x,y});
```

observe that the values of x and y have the same factor (ad − bc) in the denominator.

What if we choose a different system? Will there be a similar common denominator in the solution? Let us check.

EXAMPLE 5.2   Consider the system of linear equations

```
> eq1:=a11*x+a12*y+a13*z=f1;eq2:=a21*x+a22*y+a23*z=f2;
  eq3:=a31*x+a32*y+a33*z=f3;
```

On solving the system

```
> solve({eq1,eq2,eq3},{x,y,z});
```

notice again that the denominator %1 is common to all three variables x, y, and z. This common number is called the **determinant** of the coefficient matrix of the linear system.

Let us compute the determinant of some matrices using the Maple command det(A):

```
> A:=matrix([[a,b],[c,d]]);
> det(A);
```

The determinant of a 2 x 2 matrix is equal to the difference between the products of the diagonal and the off-diagonal elements.

The process is a bit cumbersome and tedious for matrices of larger sizes. One method for obtaining the determinant of a large matrix is referred to as the **Laplace Expansion**.

1. Select a row (column) where the expansion will occur.
2. Find the minor of each entry in the row (column) chosen in (1). The **minor** M_{ij} of the element a_{ij} is the submatrix obtained by deleting the *i*th row and the *j*th column. For the matrix

   ```
   > A:=matrix([[a11,a12,a13],[a21,a22,a23],[a31,a32,a33]]);
   ```

   The **minors** of entries a_{11}, a_{12}, a_{13} are respectively given by

   ```
   > M11:=minor(A,1,1);M12:=minor(A,1,2);M13:=minor(A,1,3);
   ```

3. Compute the **cofactor** C_{ij} of each entry in the row (column) chosen in (1). The cofactor C_{ij} of the entry a_{ij} is $(-1)^{i+j} \det(M_{ij})$;

   ```
   > C11:=(-1)^2*det(M11);C12:=(-1)^3*det(M12);
   C13:=(-1)^4*det(M13);
   ```

4. Compute the determinant as the sum of the product of each entry and its cofactor for each entry of the row chosen in (1):

   ```
   > det(A):=a11*C11 +a12*C12+a13*C13;
   > simplify(det(A));
   ```

   This is exactly the value we would get if we executed the Maple function `det(A)`.

   ```
   > det(A);
   ```

PROPERTIES OF DETERMINANTS

EXAMPLE 5.3   Determinant of the sum and of the product Consider the following matrices:

```
> A:=matrix([[2,5,7],[6,4,2],[8,4,1]]);
B:=matrix([[3,7,7],[9,6,2],[11,4,15]]);
```

Is $\det(A + B)$ equal to $\det(A) + \det(B)$?

Consider the determinant of the matrices A, B, and $C = A + B$

```
> a:=det(A); b:=det(B); c:=det(A + B);
```

Is c equal to a + b?

  Is the product of det(AB) equal to det(A) det(B)?

```
> AB:=multiply(A,B);
```

The determinant of the matrices A, B, and AB

```
> a:=det(A); b:=det(B);ab:= det(A1B);
```

Is a ∗ b equal to ab?

From this example, we deduce that in general, $\det(A + B) \neq \det(A) + \det(B)$, while **for any two n x n matrices, A and B det(AB) = det(A) det(B). (Fact 2.7a)**

EXAMPLE 5.4   **Effect of multiplying a matrix by a scalar.** Consider the matrix

```
> A:=matrix([[5,6,9,1],[2,5,7,1],[5,4,2,9],[0,7,2,6]]);
```

Multiply matrix A by the scalar k:

```
> kA:=evalm(k*A);
```

Compute and find a relation between determinant of matrix A and its scalar multiple.

```
> det(A);det(kA);
```

What do you observe? If A is a 4×4 matrix, then $\det(kA) = k^4 \det(A)$.

  What if we choose a 5×5 matrix?

```
> A:=matrix([[1,0,6,8,4],[3,5,9,10,2],[15,4,8,9,11],
[0,17,12,6,5],[2,5,7,2,1]]);
```

Multiply by the scalar k

```
> kA:=evalm(k*A);
```

Compute and find a relation between the determinant of the matrix A and its scalar multiple:

```
> det(A);det(kA);
```

What do we observe? If A is a 5×5 matrix, then $\det(kA) = k^5 \det(A)$. Can you make a general statement?

If an n x n matrix A is multiplied by a scalar k, then $\det(kA) = k^n \det(A)$. (Fact 2.7e)

EXAMPLE 5.5   **Effect of interchanging two rows.** Consider the matrix

```
> A:=matrix([[2,5,7,3],[5,4,1,8], [8,6,4,5],[9,7,4,1]]);
```

Swap two adjacent rows of matrix A

```
>  B:=swaprow(A,2,3);
```

Compute the determinants of A and B_1

```
> det(A); det(B);
```

What do you observe? Apparently det(B) = − det(A)

What if two nonadjacent rows (columns) are interchanged? Swap two nonadjacent rows of matrix A

```
> B:=swaprow(A,4,1);
> det(B);
```

What do you observe? Apparently det (B) = − det(A).

If a matrix B is obtained by interchanging any two rows of a given matrix A, then det(B) = − det(A). (Fact 2.7f)

EXAMPLE 5.6   Effect of adding a multiple of one row to another row. Consider the matrix

```
> A:=matrix([[3,5,7,9],[4,6,2,1],[7,8,5,3],[7,4,2,9]]);
```

Add 3 times row 1 to row 2

```
> B:=addrow(A,1,2,3);
```

Compute the determinants of A and B

```
> a:=det(A);b:=det(B);
```

What do you observe? Is a = b?

If a multiple of a row is added to another row, then the determinant of the matrix remains unchanged. (Fact 2.7g)

EXAMPLE 5.7   Determinant of triangular matrices. Consider two triangular matrices, one upper and the other lower:

```
> U:=matrix([[3,5,6,7,8],[0,3,2,0,5],[0,0,2,5,0],
[0,0,0,4,7],[0,0,0,0,2]]);
> L:=transpose(U);
```

Before you execute the next command, can you guess the values of the determinants of U and L? Now compute the determinant of these matrices

```
> det(U); det(L);
```

What is the relation between the determinants and the product of the diagonal elements of either U or L? Multiply the diagonal elements

```
> 3*3*2*4*2;
```

Again, can you make a general statement?

The determinant of a triangular matrix is equal to the product of its diagonal elements.

EXAMPLE 5.8   Consider the nonsingular matrix

```
> A:=matrix([[1,4,6,2],[6,3,1,7],[2,5,8,9],[7,5,1,2]]);
```

Recall that the reduced echelon form of matrix A is the identity matrix

```
> rref(A);
```

This tells us that matrix A is row equivalent to the identity matrix. Therefore, A is nonsingular. What is the determinant of A?

```
> det(A);
```

The determinant is not zero.

Now consider the matrix

```
> A:=matrix([[5,6,7,4,5],[3,2,4,7,8],[4,3,5,6,7],
  [7,5,9,13,15],[4 ,3,1,6,9]]);
```

Is matrix A row equivalent to the identity matrix?

```
> rref(A);
```

Matrix A is not row equivalent to the identity matrix. Therefore, A is singular. Again compute the determinant of A

```
> det(A);
```

Can you make a general statement?

An $n \times n$ matrix A is nonsingular if and only if det(A) is not zero. (Fact 2.8)

CRAMER'S RULE FOR SOLVING SYSTEMS OF LINEAR EQUATIONS

In Cramer's rule, a method based on determinants for solving a system of linear equations, the system must be a square system and the coefficient matrix must be nonsingular; that is, its determinant is nonzero (Facts 2.10 and Fact 2.11).

EXAMPLE 5.9   Consider the system

```
> eq1:=x-3*y+4*z=2; eq2:=-x-4*y+3*z=-2;
  eq3:=2*x-5*y+6*z= 5;
```

The coefficient matrix is

```
> C:=matrix([[1,-3,4],[-1,-4,3],[2,-5,6]]);
```

Construct the matrix A_1 obtained from the matrix C by replacing the first column of C with the right side of the system

```
> A1:=matrix([[2,-3,4],[-2,-4,3],[5,-5,6]]);
```

Find the value of the variable x

```
> x=det(A1)/det(C);
```

Construct the matrix A_2 obtained from the matrix C by replacing the second column of C with the right side of the system:

```
> A2:=matrix([[1,2,4],[-1,-2,3],[2,5,6]]);
```

Find the value of the variable y

```
> y=det(A2)/det(C);
```

Construct the matrix A_3 obtained from matrix C by replacing the third column of C with the right side of the system:

```
> A3:=matrix([[1,-3,2],[-1,-4,-2],[2,-5,5]]);
```

Find the value of the variable z

```
> z=det(A3)/det(C);
```

Note that the solution coincides with the solution obtained using

```
> solve({eq1,eq2,eq3},{x,y,z});
```

Try to create your own examples.

SUMMARY

An $n \times n$ matrix A is nonsingular (invertible)
if and only if
A is row equivalent to the identity matrix I_n
if and only if
the determinant of the matrix A is not zero
if and only if
the homogeneous system with coefficient matrix A has only the trivial solution
if and only if
the nonhomogeneous system with coefficient matrix with A has a unique solution.

EXERCISES

In the following exercises you may need to use the functions inverse, det, gausselim, rref, backsub, multiply, and add.

1. Given two 8×8 matrices A and B with det(A) $= -2$ and det(B) $= 12$, can you find the determinant of the

   a. product of the two matrices?
   b. sum A + B?
   c. inverse of A?
   d. inverse of the product of the two matrices?
   e. matrix A_1 obtained from the matrix A by interchanging two rows?
   f. matrix A_2, which is equal to 3A?

2. Apply Cramer's rule to solve the linear system

```
> eq1:= 2*x - 4*y + 7*z - w =15;
    eq2:= 3*x - 5*y + 4*z - 2* w = 13;
    eq3:= x + 4*y + 5*z - w =1;
    eq4:= 2*x + y - 7*z -8* w =12;
```

3. Given the two matrices with general entries

```
> A:=matrix([[a,b],[c,d]]); B:=matrix([[e,f],[g,h]]);
```

can you characterize all nonsingular 2×2 matrices such that $\det(A + B) = \det(A) + \det(B)$?

4. Consider the points $P_1(x_1, y_1)$, $P_2(x_2, y_2)$, and $P_3(x_3, y_3)$.

   a. Determine a condition so that the points P_1, P_2, and P_3 lie on the same line. *Hint*: Determine the slope of the line segments P_1P_2 and P_1P_3.
   b. Show that the condition in part (a) is equivalent to det(A) $= 0$, where A is the matrix

```
> A:=matrix([[1,x1,y1],[1,x2,y2],[1,x3,y3]]);
```

5. Consider the matrix

```
> A:=matrix([[1,1,1],[0,1,1],[3,-2,1]]);
```

along with the sequence of elementary matrices

```
> E1:=matrix([[1,0,0],[0,1,0],[-3,0,1]]);
> E2:=matrix([[1,0,0],[0,1,0],[0,5,1]]);
> E3:=matrix([[1,0,0],[0,1,0],[0,0,1/3]]);
> E4:=matrix([[1,-1,0],[0,1,0],[0,0,1]]);
> E5:=matrix([[1,0,0],[0,1,-1],[0,0,1]]);
```

   a. Show that the product of the matrices $E_5E_4E_3E_2E_1A$ is the identity matrix.
   b. Compute the determinant of A, using the result of part (a).

6. Repeat Exercise 5 for the matrix A but first construct the sequence of elementary matrices that reduce

```
> A:=matrix([[1,-2,0,0],[-2,1,-2,0],[0,-2,1,-2],[0,0,-2,1]]);
```

to the identity matrix.

Adjoint Method

The inverse of a nonsingular matrix A can be constructed by forming the adjoint of the matrix A.

Initialize the packages

```
> with(linalg):with(linmat);
```

The steps of the adjoint method algorithm are as follows.

1. Find the cofactor of each entry of the matrix A. The cofactor of an entry a_{ij} is defined as

$$C_{ij} = (-1)^{i+j} m_{ij}$$

   where m_{ij} is the determinant of the submatrix M_{ij}, the minor, of the entry a_{ij}.

2. Replace each entry of matrix A by its cofactor to get a new matrix C. This matrix is called the **cofactor matrix**.

3. The transpose of the matrix C is called the adjoint matrix of A and is denoted by Adj(A).

4. The inverse of A is then obtained by multiplying the matrix Adj(A) by $\frac{1}{\det(A)}$.

Let A be an $n \times n$ nosingular matrix. Then the inverse of A is given by $A^{-1} = \frac{1}{\det(A)} Adj(A)$. (Fact 2.9)

LEARNING THE PROCESS

Use the demonstration mode of the function `inverse` to learn this algorithm.

EXAMPLE 6.1 Find the inverse of the following matrix using the adjoint method

```
> A:=matrix([[1,3,5],[5,3,6],[8,4,2]]);
> inverse(A);
```

Observe that

```
> Adjoint(A):=matrix([[-18,14,3],[38,-38,19],
  [-4,20,-12]]);
```

The product of A with its adjoint is the matrix

```
> multiply(A,Adjoint(A));
```

This is the **identity matrix** multiplied by the **determinant** of the matrix A.

Repeat the procedure with as many matrices as it takes to learn the process.

EXERCISES

In the following exercises you may need to use the functions `inverse, det, gausse-lim, rref, backsub, multiply, add, minor(A,i,j), cofactor(A,i,j),` and `adjoint(A)`.

1. Consider the matrix

```
> A:=matrix([[1,-2,3],[3,1,-2],[-2,3,1]]);
```

   a. Find the minors of the entries 3, 1, and −2 in the second row.
   b. Find the cofactors of the entries 3, 1, and −2 in the second row
   c. Construct the adjoint matrix.
   d. Deduce the inverse of A.
   e. Use the interactive mode of the function `inverse` to confirm the answer to (d).

2. Consider the matrix

```
> A:=matrix(4,4):
> for i from 1 to 4 do for j from 1 to 4 do
    A[i,j]:=1/(i+j-1);od;od;print('A =', A);
```

   a. Determine the adjoint matrix B of matrix A
   b. What is the product C of matrix A and its adjoint?
   c. Perform the scalar multiplication of det(A) and C. What do you get?
   d. Write the inverse of matrix A using the result of part (c).

3. Find the matrix A whose determinant is 5 and whose adjoint matrix is

```
> B:=matrix([[1,-3,4,5],[9,5,3,0],[4,-2,1,5],[8,3,8,1]]);
```

LU-Decomposition

The LU-decomposition of a matrix A is a factorization of a matrix A into the product of upper and lower triangular matrices.

Initialize the packages

```
> with(linalg): with(linmat);
```

Let us start with a simple example of a 2×2 matrix to understand this decomposition.

EXAMPLE 7.1 Let A be the 2×2 matrix

```
> A:=matrix([[1,2],[3,4]]);
```

The Gauss elimination yields an upper triangular matrix

```
> U:=gausselim(A);
```

The matrix U has been obtained from A by multiplying row 1 of A by -3 and adding the result to row 2. Perform the same operation on the identity matrix

```
> id:=diag(1,1);
```

to get the elementary matrix E_1

```
> E1:=addrow(id,1,2,-3);
```

What is the product of E_1 and A?

```
> U:=multiply(E1,A);
```

As you see, $E_1 A = U$. Since all elementary matrices are invertible, $A = (E_1)^{-1}U$. Find the inverse of E_1

```
> L:=linalg[inverse](E1);
```

Thus the matrix L is a lower triangular matrix. If we multiply L by U we get back matrix A

```
> LU=multiply(L,U);
```

Writing A as the product of a lower and upper triangular matrices is referred to as **LU-Decomposition**.

LEARNING THE PROCESS

Invoke LUdecomp and select the demonstration mode to learn this process.

EXAMPLE 7.2   Find the LU-decomposition of the matrix A

```
> A:=matrix([[1,3,6],[4,5,7],[6,3,1]]);
> LUdecomp(A);
```

Repeat with your favorite examples to learn the procedure. When you are sure that you have learned LU-decomposition, proceed to use the interactive mode of LUdecomp.

LU-DECOMPOSITION AND SYSTEMS OF LINEAR EQUATIONS

LU-decomposition is an alternate method for solving linear systems.
The steps employed in solving the system Ax = b using the LU-Decomposition are:

1. Decompose the matrix A into a lower and upper triangular matrices L and U respectively
2. Solve the auxiliary problem Ly = b for y using a forward substitution scheme.
3. Solve the auxiliary problem Ux = y for x using a back substitution scheme.

EXAMPLE 7.3   Consider the system of linear equations

```
> eq1:= x1 - x2 + x3 =10; eq2:= 2*x1 - x2 + 3* x3 = 6;
eq3:= x1 + 3*x2 + x3 = 5;
```

The coefficient matrix, the unknown vector, and the right side are given by

```
> A:=matrix([[1,-1,1],[2,-1,3],[1,3,1]]);
> x:=matrix([[x1],[x2],[x3]]);
> b:=matrix([[10],[6],[5]]);
```

Find the LU-decomposition of matrix A by executing the nostep mode of the function

```
> LUdecomp(A);
```

Thus matrix A can be decomposed into the product of lower and upper triangular matrices

```
> L:=matrix([[1,0,0],[2,1,0],[1,4,-4]]);
> U:=matrix([[1,-1,1],[0,1,1],[0,0,1]]);
```

Introduce the unknown y

```
> y:=matrix([[y1],[y2],[y3]]);
```

to form the matrix equation Ly = b

```
> multiply(L,y)=evalm(b);
```

This is equivalent to the system

```
> y1=10; 2*y1+y2=6; y1+4*y2-4*y3=5;
```

Use **forward substitution** (solve for y_1, then for y_2, and so on)

```
> y1:=10; y2:=-14; y3:=-51/4;
```

Therefore,

```
> y:=matrix([[y1],[y2],[y3]]);
```

Solve for x using the matrix equation $Ux = y$

```
> multiply(U,x) = evalm(y);
> x1-x2+x3=10; x2+x3=-14; x3=-51/4;
```

Use **backsubstitution** to obtain the solution

```
> x1=43/3; x2=-5/4; x3=-51/4;
```

EXERCISES

In the following exercises you may need to use the automated functions LUdecomp, inverse, multiply, gausselim, rref, and backsub.

1. Consider the system of equations

```
> eq1:=2*x1+4*x2+6*x3+x4=12;
    eq2:=x1+11*x2+4*x3+7*x4=10;
    eq3:=3*x1+17*x2+13*x3+12*x4=15;
    eq4:=2*x1+18*x2+8*x3+13*x4=20;
```

a. Write the coefficient matrix A of the system.
b. Obtain the LU-decomposition of matrix A.
c. Solve the system using the decomposition in part (b).

2. Consider the matrix

```
> A:=matrix([[3,5,7,9],[2,5,7,2],[5,4,6,9],[3,8,1,2]]);
```

and the lower triangular matrix of its LU-decomposition

```
> L:=matrix([[3,0,0,0],[2,5/3,0,0],[5,-13/3,2/5,0],
    [3,3,-51,-418]]);
```

Find the upper triangular matrix for which $LU = A$.

3. Consider the matrix

```
> A:=matrix([[13,5,17,9],[-2,5,-7,2],[5,14,6,-9],
```

```
    [3,-8,11,2]]);
```

and the upper triangular matrix of its LU-decomposition

```
> U:=matrix([[1,5/13,17/13,9/13],[0,1,-19/25,44/75],
    [0,0,1,-733/324], [0,0,0,1]]);
```

Find the lower triangular matrix for which $LU = A$.

4. Does the nonsingular matrix

```
> A:=matrix([[0,1],[1,0]]);
```

have an LU-decomposition? Explain your result.

5. Does the singular matrix

```
> A:=matrix([[0,0],[1,2]]);
```

have an LU-decomposition? If there is such a decomposition, is it unique? Explain your result.

Basic Properties of Matrices

Purpose

The objective of this lab is to study the algebra of matrices and enforce their properties. Some exploratory activities are included.

Automated Linalg Functions

In this lab, you will use the automated functions `inverse`, `commute`, `trsum`, `trproduct`, `trinverse`, `transtrans`, `LUdecomp`. For help with a function, type at the Maple prompt >?function name; for example, >?LUdecomp;

INSTRUCTIONS

1. To execute a statement, move the cursor to the input line using the mouse or the key board and hit 'enter'.
2. While executing a function, create Maple input regions if needed; otherwise, the output will not appear in the desired region.
3. Execute the following commands to load the packages

   ```
   > with(linalg):with(linmat);
   ```

TASK 1

The purpose of this task is to review basic matrix operations and their properties.

Activity 1

Are there are any values for x, y, and z for which the matrix

```
> A:=matrix([[x-1,2,1-y],[6,1,0]]);
```

is equal to the matrix B?

```
> B:=matrix([[3,2,4],[x+z,1,y+z+1]]);
```

a. What are the equations that x, y, and z must satisfy for matrices A and B to be equal?

```
> eq1:= ; eq2:= ; eq3:= ; eq4:= ;
```

b. Solve the equations using the Maple command

```
> solve({eq1,eq2,eq3,eq4},{x,y,z});
```

c. What are the values of x, y, and z for which matrices A and B are equal?

Activity 2

1. Find all 2×2 matrices A and B for which $AB = BA$. Consider the two matrices A and B:

```
> A:=matrix([[a,b],[c,d]]);
> B:=matrix([[e,f],[g,h]]);
```

  a. Form the products AB and BA.
  b. Write the system of equations resulting from $AB = BA$.
  c. Solve the resulting equations.
  d. Write the set of all possible matrices A and B satisfying the property $AB = BA$.

2. Matrix A with general entries is

```
> A:=matrix([[a,b],[c,d]]);
```

Find all 2×2 matrices for which $A^2 = A$.

  a. Form the product A^2

```
> A2:=multiply(A,A);
```

  b. Write the system of equations resulting from $A^2 = A$.
  c. Solve the resulting equations.
  d. Write the set of all possible 2×2 matrices satisfying the property $A^2 = A$.
  e. Based on part (d), write examples of a 3×3 and 4×4 matrices for which $A^2 = A$.
    Note: Matrices that satisfy the relation $A^2 = A$ are called **idempotent matrices.**

3. The matrix A with general entries is

```
> A:=matrix([[a,b,c],[d,e,f],[g,h,i]]);
```

Characterize all 3×3 matrices for which matrix A is equal to its transpose. *Hint*: Compare the entries of A and its transpose.

Activity 3

Consider the matrix

```
> A:=matrix([[x,1],[0,1]]);
```

a. Compute the following powers of matrix A: A^2, A^3 and A^4.
b. Is there a pattern from which you can deduce entry $A[1,1]$ of A^n? If so, what is this entry?
c. Is there a pattern from which you can deduce entry $A[1,2]$ of A^n? If so, what is this entry?
d. Deduce a formula for A^n. Enter the formula. *Hint*: Recall the sum of the geometric progression.
e. Is there a value of x for which the matrix A^n is equal to the matrix B?

```
> B:=matrix([[-1,1],[0,1]]);
```

TASK 2

The purpose of this task is to use the interactive mode of the function `inverse` to determine the inverse of a matrix and then write the solution of a system of linear equations in terms of the inverse.
Consider the matrix

```
> A:=matrix([[-1,2,3],[4,2,8],[5,1,c]]);
```

a. Find all the values of c for which the inverse exists.
b. Choose a value of c from part (a) for which the inverse exists, and enter the new matrix A. Use the interactive mode of `inverse` to compute the inverse using (1) Gauss-Jordan method. (2) adjoint method. Show all steps.
c. Use the inverse of matrix A to obtain two solutions to the system $Ax = b$, where (1) $b = vector([-1,2,4])$ and (2) where $b = vector([0,0,0])$.
d. In general, what is the solution of a homogeneous system $Ax = 0$ if the inverse of the coefficient matrix A exists?
e. In general, what is the solution of a nonhomogeneous system $Ax = b$ if the inverse of the coefficient matrix A exists?

TASK 3

The purpose of this task is to explore and investigate some interesting properties of matrices.
Following is a **Hilbert matrix**

```
> A:=matrix(4,4):
```

```
> for i from 1 to 4 do for j from 1 to 4 do
A[i,j]:=1/(i+j-1);od;od;print('A=',A);
```

Activity 1

Find the inverse B of the Hilbert matrix A.

Activity 2

Let $Ax = b$ be a given system of linear equations with

```
> b:=matrix([[1],[2],[3],[4]]);
```

Find the solution of the system $Ax = b$.

Activity 3

Let us replace the matrix b by the matrix b_1, obtained by adding 1 to each entry of b:

```
> b1:=matrix([[2],[3],[4],[5]]);
```

Find the solution of the system $Ax = b_1$.

Activity 4

a. Find the matrix whose rows are the sum of the rows of matrix B.
b. Compute the difference between the solutions of $Ax = b_1$ and $Ax = b$. How does this difference compare to the sum of the rows of the matrix B?

Activity 5

Repeat Activities 3 and 4, replacing matrix b by the matrix b_2, obtained by adding 2 to each entry of the matrix b.

Activity 6

Generalize the result for Activity 5 by replacing the matrix b by the matrix b_n where b_n is the matrix obtained by adding n to each entry of the matrix b.
Can you prove this generalization? Is this a property of Hilbert matrices only?

Following is a **circulant matrix**:

```
> A:=matrix([[1,2,3,4,5],[5,1,2,3,4],[4,5,1,2,3],
[3,4,5,1,2],[2,3,4,5,1]]);
```

Activity 7

Is the inverse of circulant matrix A a circulant matrix?

Activity 8

Are the powers of circulant matrix A circulant?

Activity 9

Compute the product of matrix A with the matrix

```
> b:=matrix([[3],[3],[3],[3],[3]]);
```

Find a relation between the products Ab and matrix b. Find the real value of k and characterize all nonzero matrices B for which AB = kB, where B is

```
> B:=matrix([[a],[b],[c],[d],[e]]);
```

Can you generalize this result to any $n \times n$ circulant matrix?

EXTRA LAB PROBLEMS

1. Consider the matrix

```
> A:=matrix([[0.3,0.4,0.3],[0.5,0.3,0.2],[0.2,0.3,0.5]]);
```

a. Compute several powers of the matrix A.
b. Does it seem that the powers of A are converging to a matrix with equal entries? What is this matrix? Explain your answer.

2. Consider the matrices

```
> A:=matrix([[a,b],[c,d]]); B:=matrix([[e,f],[g,h]]);
```

a. Find two 2×2 nonsingular matrices A and B for which
$$\det(A + B) = \det(A) + \det(B).$$

b. Characterize all 2×2 nonsingular matrices A and B for which
$$\det(A + B) = \det(A) + \det(B).$$

3. Consider the matrix

```
> A:=matrix([[-1,2,3],[2,5,8],[3,8,11]]);
```

If A is a symmetric matrix, are the powers $A^n (n = 2, 3, \ldots.)$ symmetric? Verify your answer.

Polynomial Equations and Matrices

Purpose

The objective of this lab is to study polynomial equations involving matrices and to continue studying of the properties of matrices. Some exploratory activities are also included.

Automated Linalg functions

In this lab you will use the automated functions `inverse`, `commute`, `trsum`, `trproduct`, `trinverse`, `transtrans` and `LUdecomp`. To get help with function, type at the Maple prompt sign > ?function name; for example, >?LUdecomp;

INSTRUCTIONS

1. To execute a statement, move the cursor to the line using the mouse or the key board and press the enter key.

2. While executing a function, create Maple input regions if needed; otherwise, the output may not appear in the desired place.

3. Execute the following commands to load the packages.

   ```
   > with(linalg):with(linmat);
   ```

TASK 1

The purpose of this task is to review properties of matrices. Let A be a 5x5 matrix whose i,jth entry is $a[i,j] = \min(i,j)$ for $i,j = 1,2,3,4,5$.
  Enter matrix A

Activity 1

Obtain the reduced row echelon form of the matrix A. From the reduced row echelon form deduce the following:

a. What is the determinant of A?
b. Does matrix A have an inverse?
c. Find the inverse of matrix A. The resulting matrix is **tridiagonal.**
d. Show that the inverse of matrix A can be decomposed into the product of lower, diagonal and upper triangular matrices.
e. Can you guess the entries of the first row of matrix A^2? Compute A^2. Is its inverse a tridiagonal matrix? Can you guess the position of the zero entries (if any) in the inverse of A^3 and A^4?

Activity 2

If A is an $n \times n$ matrix whose i,jth entry is: $a[i,j] = \min(i,j)$ as in Activity 1, what can you conclude from your calculations in Activity 1 about the following?

$$\det(\mathbf{A}); \text{inverse}(\mathbf{A}); \text{transpose}(\mathbf{A}); \text{powers of } \mathbf{A}:$$

TASK 2

The purpose of this task is to study polynomial expressions involving matrices.

Activity 1

Consider the 2×2 matrix

```
> A:=matrix([[-2,0],[0,-2]]);
```

and the identity matrix

```
> Id:=diag(1,1);
```

a. Show that A is a solution to the polynomial $p(x) = x^2 - 4$,

```
> p:=x->x^2-4;
```

that is, show that $p(A) = A^2 - 4Id = 0$ where 0 is the 2×2 zero matrix.
b. Is the matrix A the only solution to the polynomial equation $p(x) = x^2 - 4$? Verify your assertion.

Activity 2

Consider the matrix

```
> A:=matrix([[1,2,0],[3,4,1],[0,-1,1]]);
```

and the identity matrix

```
> Id:=diag(1,1,1);
```

a. Is there a nontrivial second-degree polynomial

```
> p:=x-> a*x^2 +b*x + c;
```

for which $p(A) = 0$?

  1. Compute the quantity $p(A) = aA^2 + bA + cId$.
  2. Write the system of equations for which $p(A) = 0$, where 0 is the 3×3 zero matrix.
  3. Solve the resulting system for a, b, and c.
  4. What do you conclude? Is $p(x)$ the trivial(zero) polynomial?

b. Is there a nontrivial third-degree polynomial

```
> q:=x-> a*x^3 + b*x^2 +c*x +d;
```

for which $q(A) = 0$?

  1. Compute the quantity $q(A) = aA^3 + bA^2 + cA + dId$.
  2. Write the system of equations resulting from $q(A) = 0$.
  3. Solve for the coefficients a, b, c, and d. What do you conclude? Is $q(x)$ a nontrivial polynomial? Is $q(x)$ a unique polynomial?
  4. Compute $A - xId$, compute $\det(A - xId)$, and compare the results to those in part (3).

c. What relation, if any, can be deduced between the size of the matrix A and the degree of the polynomial so that the matrix A satisfies a nontrivial polynomial? a trivial polynomial?

TASK 3

The purpose of this task is to introduce a numerical scheme the Jacobi method for solving large systems of linear equations. The **Jacobi method** is performed as follows. Suppose we have a linear system $Ax = b$ in which the coefficient matrix A is invertible and all diagonal entries of A are nonzero. We can write A as a difference of two matrices, $A = M - N$, where M is a diagonal matrix with its diagonal entries equal to those of A. Then $Ax = b$ implies $(M - N)x = b$ and this yields the matrix equation $Mx = Nx + b$. If $\{x_k\}$ is a sequence for which $M\, x_{k+1} = Nx_k + b, k = 0, 1, 2 \dots$ and if the sequence x_k converges to x^*, then x^* is a solution to $Ax = b$. The kth iterate of the solution is given by $x_k = M^{-1}*N * x_{k-1} + M^{-1}b$.

Activity 1

Solve the following system using the Jacobi method.

```
> eq1:=10*x1 + x2 - x3 =18;
eq2:=x1 + 15* x2 + x3 = -12;
eq3:= - x1 + x2 + 20* x3 =17;
```

a. Write the coefficient matrix A and the right side matrix b

```
> A:=
> b:=
```

b. Write the matrices M and N where $A = M - N$

```
> M:=
> N:=
```

c. Compute the matrices $M^{-1}N$ and $M^{-1}b$

Activity 2

Start with the initial guess

```
> x(0) := matrix([[0],[0],[0]]);
```

a. Compute x_1 using the formula $x_k = M^{-1}*N * x_{k-1}+M^{-1}b$.
b. Compute the difference.
c. Compute $x_2, x_3, x_4, \ldots$ iteratively using the formula $x_{k+1} = M^{-1}*N*x_k+M^{-1}b$. Stop when the absolute value of each entry in the difference $x_{k+1} - x_k$ is less than 0.0001.

EXTRA LAB PROBLEMS

1. The purpose of this problem is to investigate any relation between the geometric series of real numbers and matrices. Recall that a geometric series of real numbers is given by

```
> s:=Sum(x^i, i =0.. infinity);
```

whose sum, if $-1 < x < 1$, is

```
> s:=sum(x^i, i =0.. infinity);
```

Consider the matrix

```
> A:=matrix([[1/2,0],[0,1/2]]);
```

Compute the infinite sum for matrix A:

```
> Sum(A^i,i=0..infinity);
```

a. What is a condition for matrices equivalent to the condition $-1 < x < 1$? Compute the magnitude of the matrix in some sense. For example, the 2-norm of matrix A is

```
> evalf(norm(A,2));
```

b. Find the matrices $A2 = A^2$, $A3 = A^3$, $A4 = A^4$.

c. From parts a and b, can you deduce a formula for An = A^n? Write the matrix An describing this formula:

d. What is the sum of the (1,1)-entries of the matrices Id, A, A2, A3, ..., An?

e. Compute the matrix Bn = Id + A + A2 + A3 + ... + An.

f. What is the limit of each entry in the matrix Bn as n approaches infinity? Write the resulting matrix B.

g. Find the matrix C = A − I, where I is the identity matrix.

h. Find the inverse C_1 of matrix C.

i. Compare matrix C_1 to matrix B in part (f). What do you conclude?

j. Does the geometric sum of the matrix obey similar behavior as for real numbers?

k. Does your conclusion hold for any 2 × 2 matrix regardless of its norm? Verify your assertion. *Hint*: Consider a matrix whose 2-norm is greater than 1.

2. Can you state a general the result for any $n \times n$ matrix?

3. Repeat Task 3 replacing matrix M with a lower triangular matrix with no zero diagonal elements. This method is referred to as the **Gauss-Seidel method**.

Airline Connection Problem

Purpose

The purpose of this application is to formulate a mathematical model of a network of connecting flights between various cities. The idea is to utilize matrix algebra to analyze the different routes between every pair of cities.

Initialize the packages

```
> with(linalg):with(linmat);
```

DESCRIPTION OF AIRLINE CONNECTION PROBLEM

This project proposes a model for determining whether there are connecting flights between every pair of cities. Consider an airline that serves eight cities:

	C_1	C_2	C_3	C_4	C_5	C_6	C_7	C_8
C_1		1						1
C_2						1		
C_3	1			1				
C_4							1	
C_5			1					
C_6				1				
C_7								1
C_8	1				1			

If there is direct service between C_i and C_j, the i,jth entry is 1; if there is no direct service between C_i and C_j, the entry i,j is 0. The matrix that represents this configuration is called an **accessibility matrix**.

Activity 1

Enter the accessibility matrix A that represents the service between the different cities.

Activity 2

a. Is it possible to get from C_2 to C_4 in exactly two flights? Draw the path between the cities. Is it possible to get from C_5 to C_1 in exactly two flights? Draw the path between the cities.

b. Is it possible to get from C_4 to C_7 in exactly two flights? Is it possible to get from C_6 to C_8 in exactly two flights? Verify.

Activity 3

Compute A^2. How do the entries of A^2 relate to your conclusions in Activity 2? What other information about the connecting flights between cities can be obtained from A^2?

Activity 4

By computing appropriate powers of A, show

a. that we can travel from C_6 to C_8 in exactly three flights

b. how many ways one can go from C_8 to C_1 in exactly three flights

c. that we can travel from C_4 to C_3 in exactly four flights.

Activity 5

In general, interpret the meaning of the resulting nonzero entries in the various powers of matrix A

Activity 6

Compute the sum of the first three powers of A. Examine the resulting matrix. Is it possible to travel between every pair of cities in three or fewer flights?

Activity 7

Is it possible to travel between every pair of cities in six or fewer flights?

Population Dynamics Objective

Purpose

The purpose of this application is to use ideas from matrix algebra to formulate and solve a mathematical model for predicting the population size of different cities.

Initialize the package

```
> with(linalg):with(linmat);
```

DESCRIPTION OF POPULATION DYNAMICS MODEL

In tracking population size in three counties, the census bureau noted that every year the probability of a person relocating from county C_1 to county C_2 is .03 and to county C_3 is .02, from county C_2 to county C_1 is 0.04 and to county C_3 is 0.03, and from county C_3 to county C_1 is 0.06 and to county C_2 is 0.04. The probability matrix follows.

	C_1	C_2	C_3
C_1	0.90	0.03	0.02
C_2	0.04	0.93	0.03
C_3	0.06	0.04	0.95

Let x_n, y_n, and z_n represent respectively the population size of counties C_1, C_2, and C_3 in year n.

Activity 1

Write the equations that represent the population of each county in year $n + 1$.

Activity 2

Represent the data from Activity 1 and the size of the population in a matrix form. This is called the **transition matrix** A.

Activity 3

If the initial population, in thousands, of C_1, C_2, and C_3 was 200, 300, and 400 respectively, estimate the population in each county at the end of year 1.

Activity 4

Show that the population size of each county in year 1 equivalent to A times the initial population size.

Activity 5

a. Estimate the population sizes after 5 years, 10 years, 15 years, and 50 years.
b. What will the population of each county be as n becomes very large?
c. Do you think the population size of each county will eventually reach a stable value from some n onward?
d. What is the sum of the population of the three counties in any year?

Activity 6

Suppose the population size in each year for the counties C_1, C_2, and C_3 diminishes respectively by fixed numbers a, b, and c, due to spread of diseases and other environmental factors.

a. Write the equations describing this model.
b. At the end of year 1, what is the population size of each county in terms of the initial population and the factors a, b, and c?
c. What values of a, b, and c will lead to the extinction of the population of each county at the end of year 1?
d. What values of a, b, and c will lead to the extinction of the population of each county at the end of year 10?

Neural Network

APPLICATION 2.3

Purpose

This application deals with an area from computer science—the main features of a neural net and a method for training the net so that it can recognize outputs of given inputs. The analysis exhibits the utility of matrix algebra.

Initialize the packages

```
> with(linalg):with(linmat);
```

DESCRIPTION OF NEURAL NETWORK

A neural net represents a simulation of the way we learn how to associate inputs with outputs. A neural net consists of a set of inputs, a set of layers, and a set of target outputs. Each layer consists of a collection of neurons. A net may have one or more layers. The input is not usually considered a layer. Each connection from an input to a neuron is associated with a weight. A neuron may also have a "bias" input whose weight may be set to 1. The output of each layer is an input to the next layer. The final output of the output layer is then compared with a target output. If the final output is "close" (say, within 10-20%) to the target output, we say that the net has been trained. Otherwise, the process continues and the weights are adjusted accordingly.

One method for adjusting the weights is the **back propagation** algorithm. We first adjust the weights of the output layer and propagate the process backward until all weights are adjusted. This process continues until the net produces the desired outputs for a given set of inputs. In this case we say that the net has been trained. A neural net models the way humans make association by reinforcing certain connections and weakening others which is achieved by increasing some weights and decreasing others.

In this application: we consider a **two-layer neural net.** The intermediate layer has three neurons and the output layer has four neurons.

The set of inputs consists of I = {bias, red, green, orange, small, average, large}. Each input is assigned either 1 or 0—1 if the property holds, otherwise 0. For example, red will be associated with [1,1,0,0,0,0,0] (the first entry 1 represents the bias and the second 1 represents red); similarly, green will be associated with [1,0,1,0,0,0,0]; and so on.

The set of target outputs consists of O = {apple, melon, cherry, tomato}. Each output is also assigned either 1 or 0 — 1 if the property holds, otherwise 0. For example, apple will be associated with [1,0,0,0]; tomato will be associated with [0,0,0,1]; and so on.

We want the net to learn the following rules:

- **Rule a.** If green and average, then apple

$$[1,0,1,0,0,1,0] \to [1,0,0,0]$$

- **Rule b.** If orange and large, then melon

$$[1,0,0,1,0,0,1] \to [0,1,0,0]$$

- **Rule c.** If green and large, then melon

$$[1,0,1,0,0,0,1] \to [0,1,0,0]$$

- **Rule d.** If red and small, then cherry

$$[1,1,0,0,1,0,0] \to [0,0,1,0]$$

- **Rule e.** If red and average, then tomato

$$[1,1,0,0,0,1,0] \to [0,0,0,1]$$

- **Rule f.** If orange and small, then cherry

$$[1,0,0,1,1,0,0] \to [0,0,1,0]$$

Each input is connected to each neuron i with an a priori associated random weight w_{ij} (i =1,2,3 and j = 1,2,3,4,5,6,7) to yield an input weight matrix. The result of multiplying this matrix with an input is referred to as the **neti** and represents the intermediate outputs to the next layer.

The neuron fires the intermediate outputs with strength based on the scaling function

```
> f:=t->1/(1+exp(-50*t));
```

(The factor 50 in the scaling function can be adjusted, and hence there may be different scaling functions.)

These intermediate outputs act now as new inputs for the output layer. Each new input is connected to each neuron i with a priori associated random weights m_{ij} (i = 1,2,3,4 and j = 1,2,3,4) to yield an output weight matrix. The result of multiplying this matrix with an intermediate output is the final output of the net. The final output is compared with the target output. The process is repeated with a different input each time adjusting the weights until the net learns the rules.

Activity 1

To train the net:

a. Set all weights w_{ij} associated with the inputs and the bias randomly to small numbers, say between -0.1 and 0.1. Call this 7×3 matrix Nb. Enter matrix Nb.

b. Set all weights m_{ij} associated with the target outputs randomly to small numbers, say between -0.1 and 0.1. Call this 4×4 matrix Mb. Enter matrix Mb.

You want to train the net to learn two rules:

- **Rule a.** If green and average, then apple.
- **Rule b.** If orange and large, then melon.

Input Rule a. It might be represented as:

```
> ga:=vector([1,0,1,0,0,1,0]);
```

Note that the first entry is 1 and is the bias. It is always held at this value.

c. Compute the weighted input (neti) by performing the product of ga and the matrix Nb:

```
> neti:=
```

d. Apply the scaling function to each component of this intermediate output:

```
> fneti:=
```

e. Report the outcome as a vector v with the first entry of v being the bias 1 and the other three entries being the result of d:

```
> v:=
```

f. Compute the weighted intermediate output by multiplying the vector v with the matrix Mb.

g. Apply the scaling function to the resulting vector.

h. Report the output for Rule a of the net in this series of transformations as a vector r:

```
> r:=
```

How does it compare with the target output oa?

```
> oa:=vector([1,0,0,0]);
```

Does the net recognize the output? Interpret your answer

Activity 2

Now adjust the weights and input Rule b, implementing the back propagation algorithm.

a. Adjust the weights m*ij* associated with the output layer according to the rule

$$\text{new } m_{ij} = \text{old } m_{ij} + 0.3 d_j v_i$$

where 0.3 is a suitable learning factor (this can be adjusted but must remain small) and

$$d_j = r_j(1 - r_j)(T_j - r_j), \ j = 1, 2, 3, 4 \text{ and } i = 1, 2, 3, 4$$

(Tj is target output and v$_j$ is the input to the output layer. Note that equilibrium occurs when $T_j = r_j$ or $d_j = 0$ so that new m_{ij} = old m_{ij}.)

b. Adjust the weights w_{ij} according to the rule

$$\text{new } \mathbf{w}_{ij} = \text{old}\mathbf{w}_{ij} + 0.3\mathbf{e}_j\mathbf{I}_i$$

where $e_j = v_j(1 - v_j) \sum_{i=1}^{5} \sum_{j=1}^{3} d_j m_{ij}$ (Note that equilibrium occurs when $oa[j] = r[j]$ so that new w_{ij} = old w_{ij}).

You may use this code for updating the matrices Mb and Nb and begin with the new input Rule b.

```
> UMb:=matrix(4,4):
> for i from 1 to 4 do for j from 1 to 4 do
UMb[i,j]:=Mb[i,j]+0.3*r[j]*(1-r[j])*
((oa[j]-r[j]))*v[i]; od;od; print(UMb)=evalm(UMb));
> UNb:=matrix(7,3):
> for i from 1 to 7 do for j from 1 to 3 do
UNb[i,j]:=Nb[i,j]+0.3*v[j]*(1-v[j])*
sum(r[k]*(1-r[k])*(oa[k]-r[k])*Mb[j,k], k=1..4)*r[i];
od;od; print(UNb=evalm(UNb));
```

Repeat Activity 1 c–h with Rule b as input and with the updated matrices UMb and UNb for Mb and Nb respectively. How does the output compare with the targeted output om?

```
> om:=vector([0,1,0,0]);
```

Do you notice any change in the values of the output that reflects better on the identification of the first rule?

Activity 3

Repeat Activity 2 alternating between the two input Rules a and b 30 times. (Write your own procedure or else use the function `trainnet`.)

```
> trainnet(30);
```

Do the outputs indicate that the net has been trained? If so, use the updated matrices UNb and UMb to confirm that the desired output is attained for a given input.

Activity 4

The idea now is to train the net to identify rules a, b, and c. Modify the program written in Activity 3 to accommodate the new situation. Run the program several times until you think the net has been trained. Does the net recognize the target output for a given input? If so, then learning has occurred.

Activity 5

What if we change the scaling function to the function

```
> f:=t->1/(1+exp(-alpha*t));
```

Study how the training of the net is affected by selecting different values of alpha.

Activity 6

Study how the training of the net is affected if the intermediate layer has four or five neurons instead of three neurons.

Activity 7

Study the effect of using a one-layer net instead of a two-layer net.

Linear Spaces

The study of linear or vector spaces is important in formulating the concepts of the course within a mathematical structure. Expressing a vector as a linear combination of a set of vectors is an important problem in finite state machines, image processing, computer graphics, and other disciplines. Studying the properties of the linear span of a set of vectors leads to the study of basis and dimension of a linear space. Moreover, the study of linear spaces provides the framework to cast these ideas in spaces other than the Euclidean spaces.

LESSON 3.1   Introduction to Linear Spaces

We have encountered sets that are closed under addition and multiplication. For example, the set of real numbers is closed under addition and multiplication. Also, the set of all $m \times n$ matrices M is closed under addition and multiplication by scalar. The set M satisfies under addition — the commutative, the associative, the existence of an identity element (zero matrix), and the additive inverse properties — and satisfies under multiplication by scalar

$$(k_1 + k_2)A = k_1A + k_2A$$

$$(k_1k_2)A = k_1(k_2A) = k_2(k_1A)$$

$$k_1(A + B) = k_1A + k_1B$$

$$1 * A = A.$$

for any scalars k_1 and k_2 and for matrices A, B.

A mathematical structure that satisfies such properties under the operations of addition and multiplication by a scalar is called a **linear space** or a **vector space**. In this unit, we shall focus on the **Euclidean spaces** $\mathbb{R}^n$.

Initialize the packages

```
> with(linalg):with(linspace);
```

VECTOR NOTATION

A vector in the Euclidean space $\mathbb{R}^n$ is an array consisting of n components. The Maple commands that may be used to represent vectors are

- v is a vector of three components

```
> v:=vector([1,5,2]);
```

- u is a vector of four equal entries (zero vector)

```
> u:=vector(4,0);
```

- u is a vector with five equal entries

    ```
    > u:=vector(5,1);
    ```

- A vector may be specified by a function f

    ```
    > f:=x->x^2;
    > w:=vector(4,f);
    ```

What do vectors represent **geometrically**? Physical quantities such as force, velocity, weight, and acceleration are examples of vectors. Vectors are characterized by point of application, direction, and "length" or "magnitude." Quantities that are only characterized by magnitude, such as speed and mass, represent scalars.

ALGEBRA OF VECTORS

Let us investigate the properties of the structure consisting of all vectors under appropriate definition of the operations of addition and multiplication by a scalar.

Equality

EXAMPLE 1.1   Consider the two vectors

```
> v1:=vector([2,x-y,z,x+z]); v2:=vector([z,-2,2*x,3]);
```

Set up the equations that will make the components of the two vectors equal

```
> eq1:=2=z ; eq2:=x-y=-2; eq3:=z=2*x;eq4:=x+z=3;
```

Solve the resulting equations

```
> solve({eq1,eq2,eq3,eq4}, {x,y,z});
```

Thus the two vectors are equal provided that x = 1, y = 3, and z = 2.

Two vectors are equal if and only if their corresponding components are equal.

Addition

EXAMPLE 1.2   Consider the two vectors in $\mathbb{R}^2$

```
> v:=vector([1,3]); u:=vector([4,5]);
```

Geometrically, the sum of the two vectors v and u can be displayed using the `graphvectadd` function

```
> graphvectadd(v,u);
```

Physically, this sum represents the resultant of two forces v and u acting on a moving particle. **Algebraically,** the sum of the two vectors v and u is obtained using

```
>  'v+u'=evalm(v+u);
```

The components of the vector $v + u$ are the sum of the components of the vectors v and u. The sum is also a vector in $\mathbb{R}^2$. Therefore,

1. The set of vectors is closed under addition

What if we add the vector u to the vector v?

```
>  'u+v'  =evalm(u+v);
```

Are the two vectors $v + u$ and $u + v$ equal?

2. The commutative property $u + v = v + u$ holds for any two vectors u and v.

Can you state a reason as to why this result is true in $\mathbb{R}^n$?

Does the **associative property** hold? Choose any three vectors v, u, and w in $\mathbb{R}^2$

```
>  v:=vector([2,3]);  u:=vector([4,1]);
w:=vector([1,5]);
```

Geometrically, the sum of the vectors v and u can be displayed using the graphvec‐tadd function

```
>  graphvectadd(v,(u,w));
```

and

```
>  graphvectadd((v,u),w);
```

From the graph, it seems that the two sums $(v + u) + w$ and $v + (u + w)$ are equal. **Algebraically,** the sums $(v + u) + w$ and $v + (u + w)$ are

```
>  '(v + u) + w' =evalm(evalm(v+u)+w);
>  'v + (u + w)'=evalm(v+evalm(u+w));
```

The two vectors $(v + u) + w$ and $v + (u + w)$ are equal.

3. The associative property $(v + u) + w = v + (u + w)$ holds.

Can you state a reason as to why this result is true in $\mathbb{R}^n$?

Add say, the vector w to the zero vector

```
>  z:=vector(2,0);
>  'w+z'=evalm(w+z);
```

What does this tell you?

4. The zero vector is an additive identity.

What if you add, say, the vector v, to the vector $(-v)$. **Geometrically,** this vector is opposite in direction to v.

```
>  'v + (-v)'  = evalm(v+(-1)*v);
```

5. (−v) is the additive inverse of vector v.

Scalar multiplication

As in the case of matrices, when we multiply a real scalar by a vector, each component of the vector is multiplied by the scalar. The result is a vector in the same direction of the given vector if the scalar is positive and in the opposite direction if the scalar is negative.

EXAMPLE 1.3   Consider the vector

```
> v:=vector([5,7]);
```

What is the effect of multiplying the vector v by a scalar $k = 3, -3, \frac{1}{2}, 1$? Geometrically, this can be displayed using the graphscalarmulti.

```
> graphscalarmulti(3,v);
> graphscalarmulti(-3,v);
> graphscalarmulti(1/2,v);
> graphscalarmulti(1,v);
```

Algebraically,

```
> 'k1*v' = evalm(k1*v);
```

Check if the **distributive properties** hold

```
> 'k2*v' =evalm(k2*v);
> 'k1*v + k2*v' =evalm(k1*v + k2*v).;
> '(k1+k2)*v'=evalm((k1+k2)*v);
```

Comparing $k_1 v + k_2 v$ and $(k_1 + k_2)v$, we have

6. $(k_1 + k_2)v = k_1v + k_2v$ for any scalars k_1, k_2 and any vector v.

Compare the following

```
> 'k1*(v + u)'=evalm(k1*(v+u));
> 'k1*v + k1*u' =evalm(k1*v+k1*u);
```

7. $k_1(v + u) = k_1v + k_1u$ for any scalar k_1 and vectors v and u.

Now check if the **associative property** holds:

```
> '(k1*k2)*v'=evalm((k1*k2)*v);
'k1*(k2*v)'=evalm(k1*(k2*v));
'k2*(k1*v)'=evalm(k2*(k1*v));
```

We conclude that

8. $(k_1 k_2)v = k_1(k_2 v) = k_2(k_1 v)$

Finally,

```
> '1*v'=evalm(1*v);
```

$$9.\ 1*v = v \text{ for any vector } v.$$

Conclusion

A **linear space** or a **vector space** over the set of real numbers must satisfy properties 1–9 with respect to addition and scalar multiplication.

OTHER EXAMPLES OF VECTOR SPACES

A vector, in a general linear space, does not always mean an element with a specified number of components. It may refer to a function if we are dealing with a space of functions. Many examples of such spaces are studied in precalculus or calculus courses. Here are some examples.

EXAMPLE 1.4 Think of the set of all polynomials of degree less or equal to n.

Is the sum of two polynomials a polynomial? Is the scalar multiple of a polynomial a polynomial? Does this set satisfy properties 1–9?

EXAMPLE 1.5 Think of the set of all continuous functions over a closed interval [a,b]. Is this an example of a vector space? Is the sum of two continuous functions a continuous function? Is the scalar multiple of a continuous function a continuous function?

EXERCISES

1. Consider the set of all polynomials of degree less than or equal to 3. Choose two such polynomials p and q with general coefficients:

```
> p:=x->a1*x^3 + b1*x^2 +c1*x + d1;
    q:=x->a2*x^3 + b2*x^2 +c2*x + d2;
```

   a. Is the sum $p + q$ a polynomial?
   b. Is the scalar multiple $k * p$ a polynomial?
   c. Is any of the nine linear space properties being violated?
   d. Is this set a linear space under the usual addition and scalar multiplication of polynomials?

2. Consider the set of all 3×3 upper triangular matrices. Choose two such matrices A and B:

```
> A:=matrix([[a,b,c],[0,f,g],[0,0,h]]);
    B:=matrix([[a1,b1,c1],[0,f1,g1],[0,0,h1]]);
```

    a. Is the sum of the two upper triangular matrices an upper triangular matrix?
    b. Is the scalar multiple of an upper triangular matrix an upper triangular matrix?
    c. Is any of the nine linear space properties being violated?
    d. Is this set a linear space under the usual addition and scalar multiplication of matrices?

3. Repeat Exercise 2 for

    a. the set of all 3×3 invertible matrices.
    b. the set of all 3×3 symmetric matrices.

Linear Combination and Spanning

A **problem** that arises in analyzing certain type of applications is to **find a minimal set that represents the underlying structure**. Examples of such applications include partitioning a set into equivalence classes, determining the minimal spanning tree in graph theory and in the study of data structures in computer science, and determining the minimal states for a finite state machine in automata theory. This type of problem motivates us to study the following:

Given a set of vectors $S = \{v_1, v_2, \ldots, v_n\}$ and a vector w, we want to check if w can be expressed as a sum of the given vectors and if the set S is a minimal set with this property.

Initialize the packages

```
> with(linalg):with(linspace);
```

LINEAR COMBINATION

EXAMPLE 2.1   Can the vector

```
> w:=vector([-4,-20,27]);
```

be written as a sum of scalar multiples of the vectors v and u?

```
> v:=vector([2,4,6]); u:=vector([-2,-8,7]);
```

That is, is $w = c_1 v + c_2 u$ for some scalars c_1 and c_2? To show this, we need to go through two steps.

   1. Evaluate $c_1 v + c_2 u$

```
> 'c1*v+c2*u'= evalm(c1*v+c2*u);
```

For the vector w to be equal to $c_1 v + c_2 u$,

```
> evalm(c1*v+c2*u)=evalm(w) ;
```

we need to check if there are values c_1 and c_2 that satisfy the set of equations

```
> eq1:=2*c1-2*c2=-4; eq2:=4*c1-8*c2=-20;
eq3:=6*c1+7*c2=27;
```

2. Proceed to solve this **nonhomogeneous** system of equations.

Form the augmented matrix of the above system

```
> AUG:=matrix([[2,-2,-4],[4,-8,-20],[6,7,27]]);
```

Apply Gauss elimination to the augmented matrix to get the echelon form

```
> AUG1:=gausselim(AUG);
```

Apply backsubstitution to AUG1 to solve the system (if consistent)

```
> [c1,c2]=backsub(AUG1);
```

Since the nonhomogeneous system of equations has a solution $c_1 = 1$ and $c_2 = 3$, the vector w can be written as a combination of the vectors v and u

```
> evalm(w)=1*evalm(v)+3*evalm(u);
```

In general,

A vector w is a linear combination of a set of vectors v_1, v_2, ..., v_n if one can find scalars c_1, c_2, ..., c_n such that

$$w = c_1 v_1 + c_2 v_2 + \ldots + c_n v_n.$$

Expressing a vector w as a linear combination of a given set of vectors reduces to solving a nonhomogeneous system of linear equations. That is,

vector w is a linear combination of the given vectors $v_1, v_2, \ldots, v_n$ if and only if the associated nonhomogeneous system is consistent. (Fact 3.3)

LEARNING THE PROCESS

Let us repeat Example 1.1 using the function lincomb. This function demonstrates step by step the process of expressing a given vector as a linear combination of a given set of vectors. Experiment with any examples you wish. Here are some examples.

EXAMPLE 2.2   Consider the vectors in $\mathbb{R}^3$

```
> v1:=vector([2,4,6]); v2:=vector([-2,-8,7]);
w:=vector([-4,-20,27]);
```

Is w a linear combination of v_1 and v_2?

```
> lincomb(v1,v2,w);
```

If w is a linear combination, in how many ways can w be expressed as a linear combination of v_1 and v_2?

In this example, we see that w is uniquely represented as a linear combination of the given vectors.

Find another vector w that is a linear combination of v_1 and v_2

```
> w :=2*evalm(v1)+(-4)*evalm(v2);
```

Describe the set of all linear combinations of the given vectors v_1 and v_2.

EXAMPLE 2.3   Consider the vectors

```
> v1:=vector([2,4,5]);v2:=vector([3,6,8]);
  v3:=vector([1,6,3]); w:=vector([4,6,2]);
```

Is w a linear combination of v_1, v_2, and v_3? If so, in how many ways can w be expressed as a linear combination?

```
> lincomb(v1,v2,v3,w);
```

In this example, w is also uniquely represented as a linear combination of the given vectors v_1, v_2, and v_3.

Find another vector w that is a linear combination of v_1, v_2, and v_3.

```
> w :=5*evalm(v1)+2*evalm(v2)+(-1)*evalm(v3);
```

Describe the set of all linear combinations of vectors v_1, v_2, and v_3.

EXAMPLE 2.4   Consider the vectors

```
> v1:=vector([2,4,5]); v2:=vector([3,6,8]);
  v3:=vector([7,14,18]); w:=vector([6,12,18]);
```

Is w a linear combination of v_1, v_2, and v_3?

```
> lincomb(v1,v2,v3,w);
```

In how many ways can the vector w be expressed as a linear combination of v_1, v_2, and v_3? In this example, vector w can be expressed in infinitely many ways as a linear combination of the given vectors.

Find a vector that is a linear combination of vectors v_1, v_2, and v_3.

EXAMPLE 2.5   Consider the vectors

```
> v1:=vector([2,4,5]);v2:=vector([3,6,8]);
  v3:=vector([7,14,18]); w:=vector([4,6,2]);
```

Is w a linear combination of v_1, v_2, and v_3?

```
> lincomb(v1,v2,v3,w);
```

Vector w cannot be expressed as a linear combination of the given vectors v_1, v_2 and v_3.

EXAMPLES OF SPANNING SETS

The set of all linear combinations S of a given set of vectors $S_1 = \{v_1, v_2, \ldots, v_n\}$ is defined by

$$S = \{c_1 v_1 + c_2 v_2 + \ldots + c_n v_n \mid c_1, c_2, \ldots, c_n \text{ are scalars}\}.$$

S is called the **span** of set S_1. We denote this by $S = \text{span}(S_1)$.

EXAMPLE 2.6   Consider the vectors

```
> v1:=vector([1,0]);v2:=vector([0,1]);
```

The set $S = \{c_1 v_1 + c_2 v_2 \mid c_1 \text{ and } c_2 \text{ are real numbers}\} = \{(c_1, c_2) \mid c_1 \text{ and } c_2 \text{ are real numbers}\}$ is equal to the space $\mathbb{R}^2$. Thus the span$(\{v_1, v_2\})$ of vectors v_1 and v_2 is the space $\mathbb{R}^2$.

EXAMPLE 2.7   Consider the vectors

```
> v1:=vector([2,4,5]);v2:=vector([3,6,8]);
v3:=vector([7,14,18]);w:=([a1,a2,a3]);
```

Is w in the span$(\{v_1, v_2, v_3\})$? That is, can we find scalars c_1, c_2 and c_3 such that

```
> evalm(w) = c1*evalm(v1)+c2*evalm(v2)+c3*evalm(v3);
```

Construct the augmented matrix associated with the system of linear equations resulting from the above vector relation

```
> AUG:=matrix([[2,3,7,a1],[4,6,14,a2], [5,8,18,a3]]);
```

Apply Gauss elimination

```
> gausselim(AUG);
```

From the echelon form of the matrix, is the system consistent for all choices of a_1, a_2, a_3? Equivalently, is the system consistent for any arbitrary vector in $\mathbb{R}^3$?

What condition(s) must be imposed on a_1, a_2, and a_3 so that the system is consistent? In this example, the system is consistent if

```
> a2-2*a1=0;
```

This condition implies that not every vector of $\mathbb{R}^3$ is a linear combination of the given vectors v_1, v_2, and v_3. Hence the vectors v_1, v_2, and v_3 do not span all $\mathbb{R}^3$. They span a subset $S = \{[a_1, a_2, a_3] \mid a_2 - 2a_1 = 0\}$ of $\mathbb{R}^3$.

EXAMPLE 2.8   Given the vectors in $\mathbb{R}^3$

```
> v1:=vector([2,4,5]);v2:=vector([3,6,8]);
v3:=vector([0,4,1]);w:=([a1,a2,a3]);
```

Is w in the span$(\{v_1, v_2, v_3\})$? That is, can we find scalars c_1, c_2, and c_3 such that

```
> evalm(w) = c1*evalm(v1)+c2*evalm(v2)+c3*evalm(v3);
```

Construct the augmented matrix associated with the system of linear equations resulting from this vector relation

```
> AUG:=matrix([[2,3,0,a1], [4,6,4,a2], [5,8,1,a3]]);
```

Apply Gauss elimination

```
> gausselim(AUG);
```

From the echelon form of the matrix, is the system consistent for any choice of a_1, a_2, a_3? or is there a condition on a_1, a_2 and a_3 under which the system is consistent?

In this example, the system is consistent regardless of the choice of a_1, a_2, or a_3. That is every vector $[a_1, a_2, a_3]$ of $\mathbb{R}^3$ can be expressed as a linear combination of v_1, v_2, and v_3. Therefore, the vectors v_1, v_2, and v_3 do span all $\mathbb{R}^3$.
A property of the linear span is:

Let $S = \{v_1, v_2, \ldots, v_n\}$ be a subset of a linear space V. The set of all linear combinations of S, L(S), is closed under addition and scalar multiplication of V. (Fact 3.2)

EXERCISES

In the following exercises you may need to use the automated function `lincomb` and the Maple commands `evalm`, `evalm(A+B)`, `matrix`, `vector`, and `solve`.

1. Express the vector

```
> w:=vector([8,1,6,7,25]);
```

as a linear combination of the vectors

```
> v1:=vector([1,0,2,- 3,4]); v2:=vector([2,3,- 4,6,7]);
    v3:=vector([4,-5,6,7,10]);
```

using the interactive mode of `lincomb`.

2. Find a vector

```
> w:=vector([a,b,c,d]);
```

that is not in the span of the vectors

```
> v1:=vector([1,1,2,3]); v2:=vector([1,0,-2,- 3]);
    v3:=vector([4,3,4,6]);
```

You may either use `lincomb` or alternatively (a) express w as a linear combination of v_1, v_2, and v_3 and (b) extract the linear equations resulting from this linear combination. Write the augmented matrix and apply an appropriate algorithm to decide on the consistency of the system: From the reduced form of the augmented matrix, determine a relation so that the system is inconsistent. Display several vectors w that are not in the span of v_1, v_2, and v_3.

3. Describe the linear span of vectors

```
> v1:=vector([1,-1,0]); v2:=vector([0,1,0]);
```

4. In Exercise 3, will the linear span of vectors v_1 and v_2 change if you add the vector

```
> v3:=vector([1,2,3]);
```

5. In Exercise 4, will the linear span of vectors v_1, v_2, and v_3 change if you we the vector

```
> v4:=vector([5,7,9]);
```

6. Does the following set {A1,A2,A3,A4} of 2×2 matrices span all 2×2 matrices with real entries?

```
> A1:=matrix([[1,-1],[1,1]]);A2:=matrix([[1,-1],[0,1]]);
    A3:=matrix([[1,-1],[1,0]]);A4:=matrix([[1,0],[1,1]]);
```

Subspaces

Which subsets of a given linear space inherit, under addition and scalar multiplication, the same properties of the linear space?

Initialize the packages

```
> with(linalg):with(linspace);
```

Before we proceed to characterize subsets of a linear space that inherit the structure of the space itself, let us review examples from the set of real numbers $\mathbb{R}$.

EXAMPLE 3.1   Consider the subset

```
>   S:= {-1,0,1};
```

Is subset S closed under multiplication? Is subset S closed under addition?

One can find subsets of the vector space $\mathbb{R}$ of real numbers that do not inherit the same properties as $\mathbb{R}$ itself. How do we characterize subsets that inherit the same properties as the space itself?

EXAMPLE 3.2   Consider the following subset of $\mathbb{R}^2$

```
>  'S1={[x,y]  |  2*x+3*y=1}';
```

Choose two vectors in S_1

```
> v:=vector([-4,3]);u:=vector([-7,5]);
```

Compute their sum

```
>  'v+u'=evalm(v+u);
```

Does vector v + u belong to subset S_1? Why?

What about the scalar multiplication? Let k be any scalar. Then

```
>  'k*v'=evalm(k*v);
```

Does vector $k*v$ belong to subset S_1 for any value of the scalar k? Why? Subset S_1 is neither closed under addition nor under scalar multiplication. Thus subset S_1 does not inherit same properties as $\mathbb{R}^2$.

Consider now the subset S_2

```
> 'S2={(x,y) | 2*x+3*y=0}';
```

Is subset S_2 closed under addition? under scalar multiplication?
Subset S_2 is closed under the two operations. Such a subset of $\mathbb{R}^2$ is called a **subspace**
of $\mathbb{R}^2$.
Let us geometrically observe the distinction between the two sets S_1 and S_2:

```
> with(plots):implicitplot({2*x+3*y=0,2*x+3*y=1},
x=-5..5,y=-5..5);
```

We observe that the graph of S_2 passes through the origin while the graph of S_1 does
not. The zero vector belongs to the set S_2 but does not belong to the set S_1.

EXAMPLE 3.3   Consider an example in $\mathbb{R}^3$

```
> 'S1={[x,y,z] | 2*x+3*y+z=0}';
> 'S2 ={[x,y,z] | 2*x+3*y+z=100}';
> implicitplot3d({2*x+3*y+z=0,2*x+3*y+z=100}
,x=-5..5,y=-5..5,z=0..100,axes=normal);
```

Graphically, can you tell which subset does not contain the zero vector?
What is the implication of not having the zero vector in the set?
Is S_2 closed under addition and scalar multiplication? Choose two vectors from S_2 and
repeat Example 3.2.

Is S_1 closed under addition and scalar multiplication? Choose two vectors from S_2
and repeat Example 3.2.
Therefore, a non-empty set S of a linear space V is a **subspace** of V if the following two
conditions are satisfied:

- If v_1 and v_2 belong to S, then $v_1 + v_2$ is in S; that is, S is closed under addition.
- If v belongs to S and k is any scalar, then $k * v$ is in S; that is, S is closed under
  scalar multiplication.

Is a linear space V a subspace of itself?
Is the set consisting of the zero vector a subspace of any linear space?
If a subset does not contain the zero element, can this subset be a subspace? **(Fact 3.1)**
What are all the subspaces of $\mathbb{R}^2$? of $\mathbb{R}^3$? In general, what are the subspaces of $\mathbb{R}^n$?

Let $\{v_1, v_2, \ldots v_n\}$ be a set of vectors in a linear space V. Is the set of all linear
combinations — that is, the span($\{v_1, v_2 \ldots v_n\}$) — a subspace of the given vector space?
(Fact 3.2)

EXAMPLE 3.4   Consider the homogeneous system of linear equations

```
> eq1:=x1+3*x2+5*x3 = 0; eq2:=3*x1+2*x2+5*x3 = 0;
```

The system is defined by matrix multiplication $A * x = 0$ where A and x are

```
> A:=matrix([[1,3,5],[3,2,5]]);
  x:=matrix([[x1],[x2],[x3]]);
```

which is equivalent to

```
> multiply(A,x)=0;
```

Upon solving the system

```
> solve({eq1,eq2},{x1,x2,x3});
```

we obtain

```
> 'S={(x1,x2,x3)|x1=(-5/7)*a,
  x2=(-10/7)*a,x3=a}';
```

Is S a subspace of $\mathbb{R}^3$?

Geometrically, S represents a line through the origin. The set S is closed under addition and scalar multiplication. Therefore, S is a subspace. Hence, the solution set of a homogeneous system is a subspace. This is called the **nullspace** of matrix A and is denoted by N(A).

Is the solution set of a nonhomogeneous system of linear equations a subspace?

LEARNING THE PROCESS

The subspace function demonstrates in a step by step manner the process of showing whether a subset is a subspace or not. Experiment with any subsets you wish. Here are some examples.

EXAMPLE 3.5   Consider the subset S given by

```
> S:={[x,y,z], x+y^2+z=0};
```

Is S a subspace of $\mathbb{R}^3$? Check using the subspace function

```
> subspace(S);
```

Which property has been violated?

EXAMPLE 3.6   Consider the subset S given by

```
> S:={[x,y,z], x+y+z=0};
```

Is S a subspace of $\mathbb{R}^3$? Again, use the function

```
> subspace(S);
```

What is the reason for your answer?

EXERCISES

In the following exercises you may need to use the automated functions `subspace` and `lincomb` and the Maple commands `evalm`, `add`, `matrix`, `vector`, `gausselim`, and `solve`.

1. Let S be a subset of $\mathbb{R}^3$ given by

```
> S:={[x,y,z], x + 2*y - 3*z =0};
```

a. Exhibit two specific elements v_1 and v_2 of the set S. Is v_1+v_2 in S? Is $k * v_1$ in S?
b. Use the `subspace` function to check whether S is a subspace or not.

```
> subspace();
```

2. The matrix A represents the augmented matrix of a homogeneous system of linear equations:

```
> A:=matrix([[1,-2,4,0],[-1,2,-4,0],[2,-4,8,0]]);
```

a. Reduce matrix A and describe the solution set S of the homogeneous system.
b. Is solution set S a subspace? Use the interactive mode of the `subspace` function.
c. Suppose you replace the augmented matrix by

```
> A:=matrix([[1,-2,4,3],[-1,2,-4,-3],[2,-4,8,6] ]);
```

d. Reduce matrix A and describe solution set S of the nonhomogeneous system.
e. Is the solution set S a subspace? Explain.

LESSON 3.4   Linear Independence

Before we find the minimal set S that spans a linear space V — that is, the smallest set S in which every vector in the linear space V can be written as a linear combination of S, — we need to introduce the notion of linear dependence.

Initialize the packages

```
> with(linalg):with(linspace);
```

EXAMPLES OF LINEAR DEPENDENT AND INDEPENDENT SETS

EXAMPLE 4.1   Consider the two vectors

```
> v:=vector([2,5]); u:=vector([4,10]);
```

Vector u is a multiple of vector v (u = 2v). This can be expressed as $u + (-2)v = 0$. That is, there are nonzero scalars $c_1 = -2$ and $c_2 = 1$ such that the combination $c_1 u + c_2 v$ is equal to the zero vector. The vectors v and u are called **dependent vectors**

Consider a linear combination of v and u $c_1 v + c_2 v$

```
> c1*evalm(v)+c2*evalm(u);
```

which is equal to the zero vector [0, 0]

```
> c1*evalm(v)+c2*evalm(u) = [0,0];
```

From this relation, we get the associated homogeneous system

```
> eq1:=2*c1+4*c2=0; eq2:=5*c1+10*c2=0;
```

Solve the system for c_1 and c_2

```
> solve({eq1,eq2},{c1,c2});
```

Does the system have a solution?
There are infinitely many solutions, of which $c_2 = 1$ and $c_1 = -2$ is one of them. This implies that there exist scalars c_1 and c_2 that are not both zero such that

```
> c1*evalm(v)+c2*evalm(u)=[0,0];
```

Geometrically, the vectors v and u are along the same line and are referred to as dependent vectors.

EXAMPLE 4.2   Now consider the two vectors

```
> v:=vector([2,5]); u:=vector([4,8]);
```

The vector u is not a multiple of the vector v. Is it possible to find nonzero scalars c_1 and c_2 for which

```
> c1*evalm(v)+c2*evalm(u) = [0,0];
```

The associated homogeneous system from this linear combination is

```
> eq1:=2*c1+4*c2=0; eq2:=5*c1+8*c2=0;
```

Solve for c_1 and c_2

```
> solve({eq1,eq2},{c1,c2});
```

Does the system have a solution? Indeed, $c_1 = 0$ and $c_2 = 0$ is the only solution. In this case

```
> c1*evalm(v)+c2*evalm(u) = [0,0];
```

implies that c_1 and c_2 are both zero.
Geometrically, the vectors v and u are not along the same line. The vectors v and u are called **independent**.

EXAMPLE 4.3   Consider three vectors in $\mathbb{R}^2$

```
> v1:=vector([2,3]); v2:=vector([5,7]);
v3:=vector([9,19]);
```

From Examples 4.1 and 4.2 we observe that dependent or independent vectors can be determined by considering the vector equation

```
> c1*evalm(v1)+c2*evalm(v2)+c3*evalm(v3)=[0,0];
```

The associated homogeneous system is

```
> eq1:=2*c1+5*c2+9*c3=0; eq2:=3*c1+7*c2+19*c3=0;
```

Solve for c_1, c_2, and c_3

```
> solve({eq1,eq2},{c1,c2,c3});
```

This system has infinitely many solutions.
Are vectors v_1, v_2, and v_3 dependent or independent? Equivalently, can we write one of the vectors v_1, v_2, or v_3 as a linear combination of the other two? Indeed, one such combination is:

```
>   [9,19] = 32*[2,3] + (-11)*[5,7];
```

That is, vectors v_1, v_2, and v_3 are dependent. In general,

A set of vectors v_1, v_2, ..., v_n of a vector space V is linearly dependent if there exist scalars c_1, c_2, ..., c_n not all zero such that

$$c_1 v_1 + c_2 v_2 + c_3 v_3 \ldots + c_n v_n = 0$$

equivalently, one of the vectors is a linear combination of the other vectors. Otherwise, the set of vectors is linearly independent; that is, any zero combination

$$c_1 v_1 + c_2 v_2 + c_3 v_3 \cdots + c_n v_n = 0$$

implies that

$$c_1 = c_2 = c_3 = \cdots = c_n = 0.$$

Determining the linear dependence or independence of a set of vectors reduces to solving a homogeneous system of linear equations. If the homogeneous system has only the trivial solution, then the set of vectors is **linearly independent**. If the homogeneous system has infinitely many solutions, the set of vectors is **linearly dependent**. (Fact 3.7 and 3.8)

To determine the **linear dependence** or **independence** of a set of vectors

1. Write the linear combination of the vectors and set it equal to zero.
2. Deduce the associated homogeneous system of linear equations.
3. Solve the homogeneous system.

LEARNING THE PROCESS

Let us consider other examples using the function `lindep` in the `linspace` package. This function demonstrates in a step by step manner the process of determining whether a set of vectors is linearly dependent or not. Experiment with any set of vectors. Here are some examples

EXAMPLE 4.4   Are the following vectors dependent or independent?

```
> v1:=vector([1,2,3]); v2:=vector([4,3,1]);
v3:=vector([5,5,4]);
> lindep(v1,v2,v3);
```

In this example the vectors are linearly dependent. In general,

If a set S of vectors is dependent, then at least one vector in S can be written as a linear combination of the others. (Fact 3.4)

EXAMPLE 4.5   Consider the vectors

```
> e1:=vector([1,0,0,0,0]); e2:=vector([0,1,0,0,0]);
e3:=vector([0,0,1,0,0]);
> e4:=vector([0,0,0,1,0]); e5:=vector([0,0,0,0,1]);
```

Can you guess whether the vectors are dependent or independent? In this example the set $\{e_1, e_2, e_3, e_4, e_5\}$ is linearly independent. In general, the set of vectors $\{e_i, i = 1, 2, 3, \ldots, n\}$, where e_i has 1 in the ith position and 0 elsewhere, is an example of an independent set.

EXAMPLE 4.6   Verify that the following set is linearly dependent:

```
> u1:=vector([0,0,0,0]);u2:=vector([2,6,7,9]);
u3:=vector([5,2,3,8]);u4:=vector([1,4,3,5]);
```

Use the nostep mode of `lindep`

```
> lindep(u1,u2,u3,u4);
```

If a set of vectors S includes the zero vector, then S is linearly dependent. (Fact 3.5)

If S is an independent set of vectors and S_1 is a subset of S, is S_1 dependent or independent?

EXAMPLE 4.7   Let S be an independent subset of $\mathbb{R}^4$ consisting of

```
> w1:=vector([1,4,2,3]);w2:=vector([-1,2,0,2]);
w3:=vector([0,0,1,1]);w4:=vector([0,0,0,1]);
```

Choose S_1 to be a subset of S

```
> S1:={w2,w3};
```

Is S_1 independent or dependent subset of S?

```
> lindep(w2,w3);
```

In general,

If S is an independent set, then any subset S_1 of S is independent. (Fact 3.6)

If S is a dependent set, is every subset of S dependent? Can we extract an independent subset S_1 of S?

EXAMPLE 4.8   Let S be a dependent subset of $\mathbb{R}^4$ consisting of

```
> v1:=vector([1,3,5]); v2:=vector([4,2,1]);
v3:=vector([0,1,1]); v4:=vector([6,8,11]);
```

and let S_1 be a subset of S

```
> S1:={v1,v3,v4};
```

Is S_1 linearly dependent or not?

```
> lindep(v1,v3,v4);
```

Set S_1 is independent. How about the set

```
> S1:={v1,v2,v4};
```

Is this set linearly dependent or not?

```
> lindep(v1,v2,v4);
```

The set S_1 is dependent.

If S is a dependent set, then not every subset of S is dependent.

EXAMPLE 4.9   Consider the set of functions

```
> f:=x->2*exp(x); g:=x->x*exp(x); h:=x->(x^2)*exp(x);
```

Are functions f, g, and h independent? Consider the relation

```
> w:=c1*f(x) + c2*g(x) + c3*h(x)=0;
```

Since this relation is valid for all x, substitute $x = 1$, $x = 2$, and $x = 0$ in w to obtain three equations to solve for c_1, c_2, and c_3

```
> eq1:=subs(x=1,w);
> eq2:=subs(x=2,w);
> eq3:=subs(x=0,w);
```

Solve the homogeneous system of equations

```
> solve({eq1,eq2,eq3},{c1,c2,c3});
```

Are the functions f, g, and h dependent or independent?

EXERCISES

In the following exercises you may need to use the automated functions `lindep`, `lincomb`, and `subspace` and the Maple commands `evalm`, `add`, `matrix`, `vector`, `gausselim`, `solve`, and `Wronskian`.

1. Use the interactive version of `lindep` to check dependence of the vectors

```
> v1:=vector([-1,2,3,4]);v2:=vector([-1,3,2,5]);
    v3:=vector([-4,9,11,17]);v4:=vector([-1,5,3,0]);
```

2. Consider the two vectors

```
> v1:=vector([-1,2,-3,4,5]); v2:=vector([1,-1,7,3,4]);
```

a. Construct a vector w that is not a linear combination of v_1 and v_2. *Hint*: Let w = vector([a, b, c, d, e], assume that w is a linear combination of v_1 and v_2, and obtain a relation or set of relations that will allow you to answer the question.

b. Are the given vectors together with the one constructed in (a) linearly independent?

   c. Can you describe a method of extending a given linearly independent set into a larger linearly independent set?

3. Consider the functions

```
> f1:=x->1; f2:=x->x; f3:=x->x^2;
```

   a. Is the set of functions linearly dependent or independent over the interval $(-1, 1)$?

   b. Construct the 3×3 matrix A whose first row is the set of functions f_1, f_2, and f_3 and whose second and third rows are the first and second derivatives of f_1, f_2, and f_3 respectively. Compute the determinant of this matrix. This is called the **Wronskian** of the functions f_1, f_2, and f_3 (type `?Wronskian` at the Maple prompt).

   c. Consider the functions

```
> g1:=x->1; g2:=x->x; g3:=x->1+x;
```

   Are the functions g_1, g_2, and g_3 linearly dependent or independent?

   d. Find the Wronskian of the functions g_1, g_2, and g_3.

   e. From parts (a)–(d) can you establish a relation between linear dependence and the Wronskian?

4. Are the functions $1, x, x^2, x^3, \ldots, x^n$ linearly independent?

Basis and Dimension

We are now ready to determine the **minimal spanning set** for a linear space.

Initialize the packages

```
> with(linalg):with(linspace);
```

EXAMPLES OF BASIS

EXAMPLE 5.1 Consider the two independent vectors

```
> e1:=vector([1,0]); e2:=vector([0,1]);
```

Is every vector in $\mathbb{R}^2$ a linear combination of the vectors e_1 and e_2?
Let w be any vector in $\mathbb{R}^2$

```
> w:= vector([x,y]);
```

If $w = [x, y]$ is a linear combination of e_1 and e_2

```
> w =c1*evalm(e1)+c2*evalm(e2);
```

then $c_1 = x$ and $c_2 = y$, hence w is

```
> evalm(w)=x*evalm(e1)+y*evalm(e2);
```

Therefore, every vector w in $\mathbb{R}^2$ is a linear combination of the linearly independent vectors e_1 and e_2. Thus the set $\{e_1, e_2\}$ is linearly independent that spans $\mathbb{R}^2$.

What if we add another vector u to the set $\{e_1, e_2\}$ to form the set $\{e_1, e_2, u\}$?

EXAMPLE 5.2 Consider the vectors

```
> e1:=vector([1,0]);e2:=vector([0,1]); u:=vector([2,5]);
```

Does the set $\{e_1, e_2, u\}$ span $\mathbb{R}^2$? Let w be any vector in $\mathbb{R}^2$

```
> w:=vector([x,y]);
```

If $w = [x, y]$ is a linear combination of e_1, e_2, and u,

```
> evalm(w)=c1*evalm(e1)+c2*evalm(e2)+c3*evalm(u);
```

This yields the system of equations

```
> eq1:=c1+2*c3=x; eq2:=c2+5*c3=y;
```

On solving the system we get

```
> solve({eq1,eq2},{c1,c2,c3});
```

Since the system is consistent, the vectors e_1, e_2, and u do span $\mathbb{R}^2$. By choosing the free variable $c_3 = 1$, the vector w can be written

```
> w=(x-2)*evalm(e1)+(y-5)evalm(e2)+evalm(u);
```

Are vectors e_1, e_2, and u linearly independent? Check using the nostep mode of `lindep`

```
> lindep(e1,e2,u);
```

We conclude that the set $\{e_1, e_2, u\}$ is linearly dependent but spans $\mathbb{R}^2$.

In Examples 5.1 and 5.2, both sets $S_1 = \{e_1, e_2\}$ and $S_2 = \{e_1, e_2, u\}$ span $\mathbb{R}^2$. However, set S_1 is linearly independent set while S_2 is linearly dependent.
Is it possible to span $\mathbb{R}^2$ using $\{e_1\}$ or $\{e_2\}$? It seems that the set $\{e_1, e_2\}$ is a minimal independent set that spans $\mathbb{R}^2$. We shall call such a set a basis for $\mathbb{R}^2$. The set $\{e_1, e_2\}$ is referred to as the **standard basis** of $\mathbb{R}^2$.

Is set $\{e_1, e_2\}$ the only linearly independent set that spans $\mathbb{R}^2$?

EXAMPLE 5.3   Consider the set of vectors

```
> v:=vector([2,7]); u:=vector([5,11]);
```

Let w be any vector in $\mathbb{R}^2$

```
> w:=vector([x,y]);
```

If w is a linear combination of v and u

```
> evalm(w)=c1*evalm(v)+c2*evalm(u);
```

this yields the system of equations

```
> eq1:=2*c1+5*c2=x; eq2:=7*c1+11*c2=y;
```

On solving the system for c_1 and c_2, we get

```
> solve({eq1,eq2},{c1,c2});
```

Since the system is consistent, vectors v and u do span $\mathbb{R}^2$.
Check if vectors v and u are independent by using the nostep mode of `lindep`

```
> lindep(v,u);
```

Any two linearly independent vectors in $\mathbb{R}^2$ will act as a minimal independent set that span $\mathbb{R}^2$.

Does a similar situation hold in $\mathbb{R}^3$?

EXAMPLE 5.4   Consider the set of linearly independent vectors

```
> e1:=vector([1,0,0]); e2:=vector([0,1,0]);
e3:=vector([0,0,1]);
```

Let w be any vector in $\mathbb{R}^3$

```
> w:=vector([x,y,z]);
```

If w is a linear combination of e_1, e_2, and e_3,

```
> evalm(w)=c1*evalm(e1)+c2*evalm(e2)+c3*evalm(e3);
```

then $c_1 = x$, $c_2 = y$, and $c_3 = z$. Hence,

```
> w = x*e1 +y*e2 +z*e3;
```

Therefore set $\{e_1, e_2, e_3\}$ is a linearly independent set that spans $\mathbb{R}^3$. The set $\{e_1, e_2, e_3\}$ is called the **standard basis** for $\mathbb{R}^3$.

What if we add one vector to the set $\{e_1, e_2, e_3\}$?

EXAMPLE 5.5   Consider the set of vectors

```
> e1:=vector([1,0,0]); e2:=vector([0,1,0]);
.e3:=vector([0,0,1]); u:=vector([2,5,6]);
```

Let w be any vector in $\mathbb{R}^3$

```
> w:=vector([x,y,z]);
```

If w is a linear combination of e_1, e_2, e_3 and u,

```
> evalm(w)=c1*evalm(e1)+c2*evalm(e2)+c3*evalm(e3)
+c4*evalm(u);
```

then we get the system of equations

```
> eq1:=c1+2*c4=x;eq2:=c2+5*c4=y;eq3:=c3+6*c4=z;
```

Upon solving this system for c_1, c_2, and c_3,

```
> solve({eq1,eq2,eq3},{c1,c2,c3,c4});
```

we see that the vectors e_1, e_2, e_3, and u do span $\mathbb{R}^3$. By choosing $c_4 = 1$, w can be written

```
> evalm(w)=(x-2)*evalm(e1)+(y-5)*evalm(e2)
+(z-6)*evalm(e3)+evalm(u);
```

Is the set $\{e_1, e_2, e_3, u\}$ linearly independent? Check using the nostep mode of lindep

```
> lindep(e1,e2,e3,u);
```

In Examples 5.4 and 5.5, both sets $S_1 = \{e_1, e_2, e_3\}$ and $S_2 = \{e_1, e_2, e_3, u\}$ span $\mathbb{R}^3$. However set S_1 is linearly independent while set S_2 is linearly dependent. Is it possible for a set of two or fewer vectors to span $\mathbb{R}^3$? It seems that set $\{e_1, e_2, e_3\}$ is a minimal independent set that spans $\mathbb{R}^3$. We shall call such a set a **basis** for $\mathbb{R}^3$.

Again, is this the only minimal independent set that spans $\mathbb{R}^3$?

EXAMPLE 5.6   Consider the vectors

```
> v1:=vector([0,1,1]); v2:=vector([1,1,2]);
v3:=vector([1,0,2]);
```

Let w be any vector in $\mathbb{R}^3$

```
> w:=vector([x,y,z]);
```

Check if w is a linear combination of the vectors v_1, v_2, and v_3

```
> evalm(w)=c1*evalm(v1)+c2*evalm(v2)+c3*evalm(v3);
```

This yields the system of equations

```
> eq1:=c2 + c3=x;eq2:=c1 + c2=y;
eq3:=c1+2*c2+2*c3=z;
```

Upon solving for c_1, c_2, and c_3

```
> solve({eq1,eq2,eq3},{c1,c2,c3});
```

we see that the vectors v_1, v_2, and v_3 do span $\mathbb{R}^3$ and vector w can be written

```
> w=(-2*x+z)*evalm(v1)+(2*x+y-z)*evalm(v2)
+(-x-y+z)*evalm(v3);
```

Is set $\{v_1, v_2, v_3\}$ linearly independent? Check using the nostep mode of `lindep`

```
> lindep(v1,v2,v3);
```

Any three linearly independent vectors in $\mathbb{R}^3$ will act as a minimal independent set that spans $\mathbb{R}^3$.

Following is a definition for the minimal spanning set.

SUMMARY

A basis B for a linear space V is a nonempty subset of V that satisfies the conditions

- B is a linearly independent set.
- B spans the linear space V.

The dimension of a vector space V, denoted by $\dim(V)$, is the number of elements in the basis. If the number is finite, then the space is called a finite dimensional space; otherwise, the space is an infinite dimensional space.

From Examples 5.1–5.4 we conclude that the elements of a minimal spanning set for a linear space V are not unique. There are many different minimal spanning sets for a linear space but the number of the elements in each set must be the same.

LEARNING THE PROCESS

Let us learn more about basis by invoking the function `basis`. You may choose your own vector sets and check whether they form basis or not. If not, examine which of the two properties fails. Here are some examples.

EXAMPLE 5.7   Do the following vectors form a basis for $\mathbb{R}^4$? If not, which property of the basis has been violated?

```
> u1:=vector([3,4,2,7]);u2:=vector([5,2,1,0]);
u3:=vector([1,3,5,6]);u4:=vector([9,9,8,13]);
```

Use the demonstration mode of `basis` to verify this

```
> basis(u1,u2,u3,u4);
```

This example shows that one of the vectors is a linear combination of the other three. Therefore, the given set is not a basis for $\mathbb{R}^4$. However a subset of the given set is a basis for a subspace of $\mathbb{R}^4$. What is the dimension of the space spanned by these vectors?

EXAMPLE 5.8   Do the following vectors form a basis for $\mathbb{R}^3$? If not, which property of the basis has been violated?

```
> u1:=vector([3,4,2]);u2:=vector([5,2,1]);
u3:=vector([1,3,5]);u4:=vector([9,9,8]);
> basis(u1,u2,u3,u4);
```

This example shows that one of the vectors is a linear combination of the other three. Therefore, the given set is not a basis for $\mathbb{R}^3$. However a subset of the given set is a basis for $\mathbb{R}^3$. What is the dimension of the space spanned by these vectors?

EXAMPLE 5.9   Do the following vectors form a basis for $\mathbb{R}^3$? If not, which property of the basis has been violated?

```
> u1:=vector([3,4,2]); u2:=vector([5,2,1]);
u3:=vector([1,3,5]);
> basis(u1,u2,u3);
```

What is the dimension of the space spanned by these vectors? Can two of the vectors form a basis for $\mathbb{R}^3$?

```
> u1:=vector([3,4,2]);u2:=vector([5,2,1]);
> basis(u1,u2);
```

In a linear space V of dimension n, how many independent vectors are needed to span the vector space V? Is it possible to have more than n vectors in a basis for V? Is it possible to have fewer than n vectors in a basis for V?

Any set of more than n vectors in $\mathbb{R}^n$ is linearly dependent and hence not a basis for $\mathbb{R}^n$. (Fact 3.9)

If a vector space is of dimension *n*, then *n* linearly independent vectors are needed to span linear space V. (Facts 3.10 and 3.11)

A basis for a linear space is not unique. All bases must have the same number of elements. (Fact 3.12)

In how many ways can the vector w be expressed as a linear combination of a basis of a linear space?

EXAMPLE 5.10   Choose a basis for $\mathbb{R}^3$

```
> v1:=vector([1,3,5]); v2:=vector([-6,4,1]);
v3:=vector([0,1,1]);
```

and a vector w

```
> w:=vector([11,-5,23]);
```

Let us express w as a linear combination of the basis

```
> lincomb(v1,v2,v3,w);
```

If B is a basis for a linear space V, then any vector w in V can be written uniquely as a linear combination of basis B. (Fact 3.13)

EXERCISES

In the following exercises you may need to use the automated functions `basis`, `lindep`, `lincomb`, and `subspace` and the Maple commands `evalm`, `add`, `matrix`, `vector`, `gausselim`, `rref`, `solve`, and `Wronskian`.

1. Consider the vectors

```
> v1:=vector([2,-3,4,5]); v2:=vector([0,1,2,3]);
    v3:=vector([3,0,1,2]); v4:=vector([4,3,2,1]);
```

  a. Verify that the set of vectors is a basis by constructing the matrix A whose rows are the given vectors and applying `rref` or `guasselim`.
  b. Verify that the set of vectors is a basis by using the interactive mode of `basis`.
  c. Express the vector

```
> w:=vector([12,-23,27,45]);
```

  as a linear combination of this basis. In how many ways can this be done?

2. Consider the vectors

   ```
   > v1:=vector([1,1,0]); v2:= vector([0,0,1]);
   ```

   a. Is the set $\{v_1, v_2\}$ a basis for $\mathbb{R}^3$?
   b. If not, extend the set $\{v_1, v_2\}$ into a basis for $\mathbb{R}^3$.
   c. Is this extension unique? If not, write another set of three vectors including v_1 and v_2 that is a basis for $\mathbb{R}^3$.

3. Find a basis for the subspace of $\mathbb{R}^4$ that is spanned by the vectors

   ```
   > v1:=vector([1,1,1,1]); v2:=vector([1,2,1,2]);
     v3:=vector([0,1,2,0]); v4:=vector([1,3,1,3]);
   ```

4. Consider the vectors

   ```
   > v1:=vector([3,k,-1,1]); v2:=vector([-1,2,k,1]);
     v3:=vector([k,4,1,2]);
   ```

   a. Determine the value of k so that the vectors span a two-dimensional space V.
   b. Find a basis for V.

5. Consider the matrices

   ```
   > A1:=matrix([[1,0],[0,0]]); A2:=matrix([[0,1],[0,0]]);
     A3:=matrix([[0,0],[1,0]]); A4:=matrix([[0,0],[0,1]]);
   ```

   a. Do the matrices form a basis for the space of all 2×2 matrices?
   b. What is the dimension of the space of all 2×2 matrices?
   c. Answer parts (a) and (b) for the space of 3×3 matrices and 4x4 matrices, and then state a result for the space of all $n \times n$ matrices.

Row Space and Column Space

The echelon form or the reduced echelon form of an augmented matrix associated with a system of linear equations indicates whether the system is consistent or not. We have so far collected the following facts concerning the consistency of systems of linear equations.

Homogeneous systems of linear equations (HS), $Ax = 0$ are always consistent. Furthermore,

1. A system of n equations in n unknowns has a unique solution (trivial solution)

    - if and only if A is row equivalent to the identity matrix.
    - if and only if the matrix A is invertible.
    - if and only if the det(A) is not equal to zero.

2. A system of n equations in m unknowns (n < m) always has infinitely many solutions.

3. A system of n equations in m unknowns (n > m) has the trivial solution or infinitely many solutions.

Nonhomogeneous systems of linear equations (NHS), $Ax = b$, are not always consistent

1. A system of n equations in n unknowns has a unique solution

    - if and only if A is row equivalent to the identity matrix.
    - if and only if the matrix A is invertible.
    - if and only if det(A) is not equal to zero.

2. A system of n equations in m unknowns (n < m or n > m) may or may not be consistent.

A reason for studying the **row space** and the **column space** of a matrix is to address the question of consistency of the (NHS).

Initialize the packages

```
> with(linalg):with(linspace);
```

Examples of Row and Column Spaces

The **row space** of an $n \times m$ matrix A is the space spanned by the rows of matrix A. Since the span of a set of vectors is a subspace, the row space of an $n \times m$ matrix A is a subspace of R^m (Fact 3.14). The number of the nonzero rows in the echelon or reduced row echelon form of A is called the **row rank** of A and is denoted by rank(A).

The **column space** of an $n \times m$ matrix A is the space spanned by the columns of matrix A. Since the span of a set of vectors is a subspace, the column space of an $n \times m$ matrix A is a subspace of R^n (Fact 3.14). The number of the nonzero columns in the echelon or reduced row echelon form of A is called the **column rank** of A and is denoted also by rank(A).

EXAMPLE 6.1   Consider the matrix

```
> A:=matrix([[1,2,3],[6,-3,-8],[8,1,-2]]);
```

Apply Gauss elimination to find the row echelon form of A

```
> B:=gausselim(A);
```

In this example, since the number of nonzero rows of the reduced matrix is 2, the row space of A is of dimension 2 or the rank(A) = 2. A basis for the row space consists of the nonzero rows of the reduced matrix B:

```
>   v1:=row(B,1);v2:=row(B,2);
```

EXAMPLE 6.2   Determine the rank and a basis for the row space of the matrix

```
> A:=matrix([[1,8,3,5],[6,-3,7,11],
  [8,1,-2,10],[14,-2,5,21]]);
```

Apply Gauss elimination to find echelon form of A

```
> B:=gausselim(A);
```

The row space is of dimension 3 — that is, rank(A) = 3 — and a basis for the row space is

```
> v1:=row(B,1) ; v2:=row(B,2);
  v3:=row(B,3);
```

EXAMPLE 6.3   Determine the column rank and a basis for the column space of the matrix

```
> A:=matrix([[1,8,3,5],[6,-3,7,11],
  [8,1,-2,10],[14,-2,5,21]]);
```

Take the transpose of A

```
> B:=transpose(A);
```

Apply Gauss elimination to the matrix B

```
> B1:=gausselim(B);
```

Transposing the result, we obtain

```
> A1:=transpose(B1);
```

The matrix A_1 implies that the column space of A, has dimension 3 ((rank(A) = 3) and its basis consists of the three nonzero column vectors in the reduced form A_1.

The dimension of the row space is equal to the dimension of the column space of a matrix A and this common number is rank(A). (Fact 3.16)

EXAMPLE 6.4   Determine the rank and a basis for the row space of the matrix

```
> A:=matrix([[1,8,3,4,12],[6,-3,7,12,15],
[-8,3,-2,10,18],[-7,11,1,14,31]]);
```

Apply gausselim to find row echelon form of A

```
> B:=gausselim(A);
```

From this we deduce that the dimension of the row space is 4 — that is, rank(A) = 4 — and a basis for the row space is

```
> v1:=row(B,1);v2:=row(B,2); v3:=row(B,3);v4:=row(B,4);
```

If A is the augmented matrix of a system of linear equations, is the system consistent?

The form of the matrix B indicates that the system is not consistent. Let us relate this to the rank of the augmented matrix and the rank of the coefficient matrix. The rank of the coefficient matrix is 3 while the rank of the augmented matrix is 4.

What if we slightly change matrix A in Example 6.4 and replace the (4,5)-entry with 30?

EXAMPLE 6.5   Consider the matrix

```
> A:=matrix([[1,8,3,4,12],[6,-3,7,12,15],
[-8,3,-2,10,18],[-7,11,1,14,30]]);
```

Apply Gauss elimination to find echelon form of A

```
> B:=gausselim(A);
```

We deduce that the row space is of dimension 3 — that is, rank(A) = 3 — and a basis for the row space is

```
> v1:=row(B,1); v2:=row(B,2); v3:=row(B,3);
```

If A is the augmented matrix of a system of linear equations, is the system consistent?

The form of matrix B indicates that the system is consistent. Let us relate this to the rank of the augmented matrix and the rank of the coefficient matrix. The rank of the coefficient matrix is 3 and the rank of the augmented matrix is 3.

EXAMPLE 6.6 Let AUG be the augmented matrix of an NHS of 4 equations with 4 unknowns:

```
> AUG:=matrix([[1,-4,2,6,11],[4,5,-3,-7,21],
[8,2,-5,6,31],[3,2,1,7,15]]);
```

Let A be the associated coefficient matrix

```
> A:=matrix([[1,-4,2,6],[4,5,-3,-7],
[8,2,-5,6],[3,2,1,7]]);
```

Compute and compare the reduced echelon forms of matrices AUG and A

```
> A1:=rref(AUG);
> B1:=rref(A);
```

Compare the number of nonzero rows of the reduced matrix B_1 and the nonzero rows of the reduced matrix A_1. Are these numbers the same? Matrix B_1 has the same number of nonzero rows as matrix A_1. From reduced matrix A_1, can you infer that the system associated with the matrix AUG is consistent? This system is consistent and has a unique solution.

EXAMPLE 6.7 Consider now the augmented matrix

```
> AUG:=matrix([[1,-4,2,6,11],[4,5,-3,-7,21],
[8,2,-5,6,31],[12,7,-8,-1,52]]);
```

and its coefficient matrix

```
> A:=matrix([[1,-4,2,6],[4,5,-3,-7],
[8,2,-5,6],[12,7,-8,-1]]);
```

Let us compute and compare the reduced echelon form of AUG and A

```
> A1:=rref(AUG);
> B1:=rref(A);
```

Compare the number of nonzero rows of reduced matrix B_1 and the nonzero rows of matrix A_1. Are these numbers the same? Matrix B_1 has the same number of nonzero rows as the matrix A_1.
From reduced matrix A_1, can you infer that the system associated with the matrix AUG is consistent? This system is consistent and has infinitely many solutions.

EXAMPLE 6.8 Consider now the augmented matrix

```
> AUG:=matrix([[1,-4,2,6,11],[4,5,-3,-7,21],
[8,2,-5,6,31],[12,7,-8,-1,12]]);
```

and the coefficient matrix

```
> A:=matrix([[1,-4,2,6],[4,5,-3,-7],
[8,2,-5,6],[12,7,-8,-1]]);
```

Compute and compare the reduced echelon form of AUG and A

```
> A1:=rref(AUG);
> B1:=rref(A);
```

Compare the number of nonzero rows of reduced matrix B_1 and the nonzero rows of reduced matrix A_1. Are these numbers the same? In this case, matrix B_1 does not have the same number of nonzero rows as matrix A_1.

From reduced matrix A_1, can you imply that the system associated with matrix AUG is consistent? This system is inconsistent.

Let us analyze systems where the number of equations is not equal to the number of unknowns.

EXAMPLE 6.9   Consider the system of equations whose augmented matrix and coefficient matrix are

```
> AUG:=matrix([[0,-4,2,8,11],[4,9,-3,-11,21],
[3,2,-5,7,31]]);
> A:=matrix([[0,-4,2,8],[4,9,-3,-11],
[3,2,-5,7]]);
```

As in Example 6.5, look at the reduced echelon form of AUG and A   .

```
> A1:=rref(AUG);
> B1:=rref(A);
```

What does the reduced matrix A_1 imply about the system of linear equations? The system has infinitely many solutions. Does the matrix B_1 have the same number of nonzero rows as reduced matrix A_1?

Coefficient matrix A has the same number of nonzero rows as augmented matrix AUG.

Consider the matrices

```
> AUG:=matrix([[0,-4,2,8,11],[4,9,-3,-11,21],
[4,5,-1,-3,21]]);
> A:=matrix([[0,-4,2,8],[4,9,-3,-11],
[4,5,-1,-3]]);
```

Look at the reduced echelon form of AUG and A:

```
> A1:=rref(AUG);
> B1:=rref(A);
```

What does reduced matrix B_1 imply about the system of linear equations? The system has no solution.

Does matrix B_1 have the same number of nonzero rows as reduced matrix A_1? The number of nonzero rows of the coefficient matrix is not equal to the number of nonzero rows of the augmented matrix.

The above examples show that there is an intimate relation between the number of independent rows of the coefficient matrix, the augmented matrix of an underlying system of linear equations and the solvability of the system.

SUMMARY

Homogeneous systems, Ax = 0 are always consistent. Furthermore

1. A system of n equations in n unknowns has a unique solution (zero solution)

   - if and only if A is row equivalent to the identity matrix.
   - if and only if the matrix A is invertible.
   - if and only if the det(A) is not equal to zero.
   - if and only if rank(A)=n.
   - if and only if A has n independent rows (columns).

2. A system of n equations in m unknowns ($n < m$) always has infinitely many solutions.

3. A system of n equations in m unknowns ($n > m$) has either a unique solution or infinitely many solutions depending on the rank(A). If rank(A) = m, then the system has a unique solution if rank(A) < m, then the system has infinitely many solutions.

Nonhomogeneous systems, Ax = b.

1. A system of n equations in n unknowns has a unique solution

   - if and only if A is row equivalent to the identity matrix.
   - if and only if the matrix A is invertible.
   - if and only if the det(A) is not equal to zero.
   - if and only if rank(A) = n.
   - if and only if A has n independent rows (columns).

2. n equations in m unknowns ($n < m$ or $n > m$) is consistent if the rank of the augmented matrix is equal to the rank of the coefficient matrix. The system is inconsistent if the rank of the augmented matrix is not equal to the rank of the coefficient matrix.

EXERCISES

In the following exercises you may need to use the automated functions `basis`, `lindep`, `lincomb`, and `subspace`

1. Consider the matrix

```
> A:=matrix([[1,2,3,4,5],[6,7,8,9,10],
     [11,12,13,14,1]]);
```

a. Obtain the echelon form of matrix A.

b. What is the dimension of the row space? Give a basis for the row space. Compare the answer to the output of Maple using

```
> rowspace(A);
```

c. Obtain the echelon form of the transpose of A.

d. What is the dimension of the column space? Give a basis for the column space. Compare the answer to the output of Maple using

```
> colspace(A);
```

e. Is the system of equations associated with augmented matrix A solvable? If so, describe the solution set.

f. Change only one entry in matrix A so that the resulting system of equations is solvable.

2. Find the value of k such that rank(A) = 2, where A is the matrix

```
> A:=matrix([[3,1,k],[k,-2,3],[2,k,2]]);
```

Basic Properties of Linear Spaces

Purpose

The purpose of this lab is to enforce basic concepts of finite dimensional spaces, with emphasis on subspaces.

Automated Linalg functions

In this lab you will use the interactive automated function `subspace`. To get help on any function type at the Maple prompt sign >?function name; for example, >?lindep;

INSTRUCTIONS

1. To execute a statement, move the cursor to the line using the mouse or the keyboard and press the enter key.
2. While executing a function, create Maple input regions if needed; otherwise, the output will not appear in the desired place.
3. Execute the following commands to load the packages

   ```
   > with(linalg):with(linspace);
   ```

TASK 1

The purpose of this task is to study some basic properties of linear spaces. Consider two vectors in $\mathbb{R}^2$.

```
> v1 := vector([a1,b1]); v2:=vector([a2,b2]);
```

Activity 1

Is the sum w of the two vectors v_1 and v_2 in $\mathbb{R}^2$? How do you get the components of vector w from the vectors v_1 and v_2?

Activity 2

Compute the vector kv_1 for any scalar k. Is vector kv_1 in $\mathbb{R}^2$?

Activity 3

Do similar conclusions for vectors in $\mathbb{R}^3$ hold as for vectors in $\mathbb{R}^2$ in Activity 1 and Activity 2? Describe the conclusions in $\mathbb{R}^n$.

TASK 2

Task 1 shows that $\mathbb{R}^n$ is closed under addition and scalar multiplication. Does every subset of $\mathbb{R}^n$ enjoy these properties?

Activity 1

Consider two subsets of $\mathbb{R}^2$:

```
> S1:={[x,y], x^2+y=0}; S2:={[x,y],x+y=1};
```

a. Does the zero vector belong to set S_2? Is S_2 a subspace of $\mathbb{R}^2$?
b. Does the zero vector belong to set S_1? Is S_1 a subspace of $\mathbb{R}^2$?
c. Invoke the automated function `subspace` to conform your answer to (b).
d. Suggest a modification in the definition of the set S_2 to make the set a subspace of $\mathbb{R}^2$. Verify your assertion using the `interactive` mode of `subspace`.

Activity 2

a. Which one of the following two sets is a subspace of $\mathbb{R}^3$?

   i. $S3 = \{[x1, x2, x3] | x_1 - x_2 + x_3 = 0\}$
   ii. $S4 = \{[x1, x2, x3] | x_1 - x_2 + x_3 = -5\}$

b. Does the zero vector belong to set S_4? Is S_4 a subspace of $\mathbb{R}^3$? Why?
c. Does the zero vector belong to set S_3?
d. Can you imply directly that S_3 is a subspace?
e. Invoke the function `subspace` to check whether S_3 is a subspace.
   What properties must be satisfied by a subset of a linear space to be a subspace?

TASK 3

The purpose of this task is to explore the relation between subspaces and the solution set of a homogeneous system of linear equations.

   Consider the homogeneous system of linear equations

```
> eq1:=x1+x2+3*x3 = 0; eq2:=2*x1-7*x2+x3 = 0;
```

Activity 1

Write the augmented matrix AUG of the system.

Activity 2

Apply Gauss elimination to reduce the matrix AUG to its row echelon form AUG1.

Activity 3

From the matrix AUG1, write down the solution set W of the system.

Activity 4

Is W a subspace? Verify your answer.

Activity 5

Describe the intersection **S** of the two sets

$$S1 = \{[x1, x2, x3] \mid x_1 + x_2 + 3x_3 = 0\}$$

and

$$S2 = \{[x1, x2, x3] \mid 2x_1 - 7x_2 + x_3 = 0\}$$

How does S relate to W? Is the intersection of two subspaces a subspace?

Activity 6

Describe the union U of the two subspaces S_1 and S_2. Take any two elements e_1 and e_2 in U. Is the sum of e_1 and e_2 in U? Is U a subspace?

EXTRA LAB PROBLEM

Repeat Activities 1 through 6 of Task 3 for the following system:

$$x_1 + x_2 - 3x_3 + x_4 = 0$$
$$2x_1 - x_3 + 2x_4 = 0$$
$$x_1 - x_2 + x_3 - x_4 = 0$$

a. Is the intersection of subspaces a subspace?
b. Is the union of subspaces a subspace?
c. The union of subspaces may not be a subspace. Investigate conditions under which the union of two subspaces S_1 and S_2 is a subspace. *Hint*: Consider cases when S_1 is a subset of S_2 or S_2 is a subset of S_1.

Linear Combination and Independence

Purpose

The purpose of this lab is to enforce the basic concepts of linear combination and linear dependence and their relation to matrices and systems of linear equations.

Automated Linalg functions

In this lab you will use the interactive automated functions `lincomb`, `lindep`, `basis`, and `subspace`. To get help with a function, type at the Maple prompt sign >?function name; for example, >?lindep;

INSTRUCTIONS

1. To execute a statement, move the cursor to the line using the mouse or the key board and press the enter key.
2. While executing a function, create Maple input regions if needed; otherwise, the output will not appear in the desired place.
3. Execute the following commands to load the packages

   ```
   > with(linalg):with(linspace);
   ```

TASK 1

The purpose of this task is to enhance your knowledge of the notion of linear combinations. Consider the set of vectors

```
> v1:=vector([1,2,3,4]); v2:=vector([4,2,1,5]);
v3:=vector([3,5,1,7]); v4:=vector([2,1,4,0]);
w:=vector([11,9,3,17]);
```

Activity 1

Is w a linear combination of vectors v_1, v_2, v_3, and v_4? That is,

```
> w:=c1*v1+c2*v2 +c3*v3 +c4*v4;
```

For w to be a linear combination of the vectors v_1, v_2, v_3, and v_4, what is the system of linear equations that will result from this linear combination? Write the system.

Is this system consistent? Verify your answer. If the system is consistent, solve for the scalars c_1, c_2, c_3, and c_4.

What do you conclude? Is w a linear combination? Write this linear combination. Use the interactive mode of function lincomb to answer Activity 1.

```
> lincomb(v1,v2,v3,v4,w);
```

Activity 2

What conditions must be imposed on the components of the vector

```
> w:=vector([a,b,c]);
```

so that it is in the span of the vectors

```
> v1:=vector([-1,2,4]); v2:=vector([3,5,6]);
v3:=vector([-1,13,22]);
```

Express w as a linear combination of v_1, v_2, and v_3. Write the associated linear equations.
Is the resulting system consistent? Verify.
What is the relation that a, b, and c must satisfy so that w is in the span of v_1, v_2, and v_3?
What is the relation that a, b, and c must satisfy so that w is not in the span of v_1, v_2, and v_3?

TASK 2

The purpose of this task is to check whether a given set of vectors is linearly dependent or not. Consider the vectors

```
> v1:=vector([1,-2,-1,1]);v2:=vector([1,-1,0,1]);
v3:=vector([3,-5,-2,3]);v4:=vector([3,2,-1,2]);
```

Activity 1

Are the vectors linearly dependent? Write the linear combination of vectors v_1, v_2, v_3, and v_4

1. What is the system of homogeneous equations associated with the above combination? Write the system.

2. Is the system consistent? If so, does the system have a unique solution or infinitely many solutions? Apply an appropriate algorithm to answer these questions.

3. What is your conclusion? Are the vectors dependent or independent?

4. If the vectors are linearly dependent, write one of the vectors as a linear combination of the other vectors. Describe the largest subspace spanned by vectors v_1, v_2, v_3, and v_4. Execute the interactive mode of lindep:

```
> lindep(v1,v2,v3,v4);
```

Activity 2

Repeat Activity 1 of Task 2 for the vectors

```
> v1:=vector([1,1,0,1,0]); v2:=vector([2,3,1,0,5]);
v3:=vector([8,0,2,3,1]);
```

EXTRA LAB PROBLEM

The vectors

```
> v1:=vector([1,2,3]); v2:=vector([-1,2,3]);
v3:=vector([0,1,-1]);
```

are linearly independent. Is the set of vectors $S = \{v_1, 2*v_1 + v_2, v_3\}$ linearly dependent or independent? Verify your assertion.

Choose a different set of linearly independent vectors and construct a set similar to set S. Is set S independent or dependent? State, if you can, a general result along the same line.

Basis of a Linear Space

Purpose

The purpose of this lab is to reinforce our knowledge of basis of a vector space and its relation to systems of linear equations.

Automated Linalg functions

In this lab you will use the automated functions `lincomb`, `lindep`, `basis`, and `subspace`. To get help with function, type at the Maple prompt sign >?function name; for example, >?lindep;

INSTRUCTIONS

1. To execute a statement, move the cursor to the line using the mouse or the keyboard and press the enter key.
2. While executing a function, create Maple input regions if needed; otherwise, the output will not appear in the desired place.
3. Execute the following commands to load the packages

   ```
   > with(linalg):with(linspace);
   ```

TASK 1

The purpose of this task is to show whether a given set of vectors is a basis for some linear space. Consider the set of vectors

```
> v1:=vector([-1,2,4,7]); v2:=vector([3,6,0,1]);
v3:=vector([1,0,9,5]); v4:=vector([3,5,1,0]);
```

Activity 1

Does this set of vectors form a basis for $\mathbb{R}^4$? Check whether the set of vectors is independent:

```
> lindep(v1,v2,v3,v4);
```

Check whether every vector w in $\mathbb{R}^4$

```
> w:=vector([a,b,c,d]);
```

is a linear combination of v_1, v_2, v_3, and v_4. Use the nostep mode of `lincomb`.

```
> lincomb(v1,v2,v3,v4,w);
```

What is your conclusion? Do the vectors form a basis?

Activity 2

Use the interactive mode of `basis` to confirm the result in Activity 1

```
> basis(v1,v2,v3,v4);
```

Activity 3

What is the dimension of the subspace spanned by vectors v_1, v_2, v_3, and v_4?

Activity 4

If you add the vector v_5 to the set of vectors v_1, v_2, v_3, and v_4

```
> v5:=vector([3,6,7,9]);
```

do vectors v_1, v_2, v_3, v_4, and v_5 form a basis? Explain.

Activity 5

If you select $n + 1$ vectors in a linear space whose dimension is n, can the $n + 1$ vectors form a basis? Explain.

TASK 2

The purpose of this task is to construct a basis for a linear space from a given set of vectors. Consider the set of vectors

```
> v1:=vector([1,3,5,-1]);v2:=vector([4,6,7,2]);
  v3:=vector([-6,4,-1,0]);v4:=vector([6,4,3,-9]);
  v5:=vector([7,-4,2,5]);v6:=vector([5,9,12,1]);
```

Activity 1

Can these vectors form a basis for $\mathbb{R}^4$? for a subspace of $\mathbb{R}^4$?

Activity 2

If not, how do you extract a basis?

EXTRA LAB PROBLEM

Extend the following set of vectors to a basis for $\mathbb{R}^4$. Is the extension unique?

```
> v1:=vector([1,0,0,1]); v2:=vector([0,1,1,0]);
```

Oil Refinery Revisited

Purpose

The purpose of this application is to use ideas about linear spaces and revisit the oil refinery project of Unit 1.

Initialize the packages

```
> with(linalg):with(linspace);
```

DESCRIPTION OF OIL REFINERY MODEL

A company runs three oil refineries R_1, R_2, and R_3. Each R_i ($i = 1, 2, 3$) produces heating oil (HO), diesel oil (DO), and gasoline (G). Suppose that one barrel of crude oil produces the following number of gallons of heating oil, diesel oil, and gasoline in each refinery:

	R_1	R_2	R_3	Demand
HO	20	5	15	550
DO	10	3	8	300
G	10	10	c	d

Assume that the number of gallons of gasoline produced from one barrel of crude oil is not known for refinery R_3 and that the total demand for gasoline is also not known.

Activity 1

Let x_i be the number of barrels (in thousands) of crude oil processed by refinery R_i($i = 1, 2, 3$). Write the system of linear equations that describes the number of barrels needed by each refinery to meet the required demand.

Activity 2

The company found out that the demand and the number of gallons of gasoline produced by each refinery are a linear combination of the other two products, heating oil and diesel oil, and their demand. Determine the values for c and d that describe the company's findings. In particular, find the number of gallons of gasoline produced by refinery R_3 and the demand level of gasoline.

Activity 3

From your findings in Activity 2, describe the range of values of the number of barrels of crude oil needed by each refinery to meet the demands.

Activity 4

If the demand for gasoline is 700, what is the minimum integral value of c that will yield a feasible integral values for x_1, x_2, and x_3?

Activity 5

After a careful study of the original table, the company decided to implement a new policy in which the number of gallons of gasoline produced by each refinery and the demand for gasoline is twice that of diesel oil. Suppose that we are given a solution $x_1 = 15$, $x_2 = 50$, and $x_3 = 0$. Find another solution if $x_3 = 20$. *Hint*: Compute the nullspace of the matrix and use this information to determine the required solution. What is an interpretation of this solution and how does it reflect on the policy?

Activity 6

The company decided to shut down refinery R_3 and at the same time maintain the demand level:

	R_1	R_2	R_3	Demand
HO	20	5	0	550
DO	10	3	0	300
G	10	10	0	600

a. Write the system of linear equations that reflects this policy.
b. Does a solution exist?
c. If there is no solution to the system $Ax = b$, we can obtain an approximate solution as follows. Premultiply the system by A^T; that is, $A^T Ax = A^T b$. If $A^T A$ is nonsingular you can find a solution to the new system using $x = (A^T A)^{-1} A^T b$. If $A^T A$ is singular, then Gauss elimination can be used to find a solution.

    1. Write the coefficient matrix A and the demand vector b of the system in (a).
    2. Find an approximate solution b_1 using the method described above. That is, find a solution x_1 using the recipe $(A^T A)^{-1} A^T b$.
    3. Compare the demand vector b with the product $b_1 = Ax_1$.
    4. Compute the absolute value of the difference $b_1 - b$ and interpret the answer.
The process described in (c) is referred to as the **least squares method**.

Tournament Matrices

Purpose

The purpose of this application is to write a mathematical model for a round-robin tournament. The idea is to utilize linear space concepts to manipulate properties of tournament matrices.

Initialize the package

```
> with(linalg):with(linspace);
```

DESCRIPTION OF TOURNAMENT MATRICES

A round-robin tournament among n players in which no match ends in a draw consists of $\frac{n(n-1)}{2}$ matches. The results are recorded in an $n \times n$ tournament matrix $A = (a_{ij})$ with $a_{ij} = 1$ if player i defeats player j, otherwise, $a_{ij} = 0$. The score s_i is the number of players defeated by player i, and the joint score s_{ij} is the number of players defeated by both players i and j.

Activity 1

Consider a tournament with three players. Write down a few 3×3 tournament matrices. How many 3×3 matrices are possible? *Hint*: Count the number of entries below the diagonal and consider the possibilities of each entry.

Activity 2

a. Is the set of all 3×3 tournament matrices linearly independent? Verify your assertion.
b. Does the set of all 3×3 tournament matrices form a basis for the set of all 3×3 matrices? If not, propose a modification in the set to render it a basis for the set of all 3×3 matrices.

Activity 3

a. Is the product of two tournament matrices a tournament matrix?

b. Are the powers of a tournament matrix also a tournament matrix? Verify your answer. Give an interpretation for the entries of different powers.

Activity 4

a. Show that the sum of A, A^T and the corresponding identity matrix is the 3×3 matrix whose entries are all 1's.
b. Compute the product of A and A^T. Explain the meaning of the entries of the resulting matrix as they relate to the score s_i and the joint score s_{ij}.
c. Show that the rank of all the 3×3 tournament matrices is always greater or equal to 1. What is the nullity of these matrices?
d. Can you find a tournament matrix A for which the product of A and A^T is equal to the identity matrix? What is the rank of this matrix?

Activity 5

Repeat Activities 1–4 for 4×4 tournament matrices.

Activity 6

A tournament matrix is said to be regular if each player has the same score. Give examples of a 3×3 and 4×4 regular matrices.

Activity 7

A tournament matrix is said to be doubly regular provided the joint scores are all equal to some number m. Give examples of 3×3 and 4×4 such matrices.

Inner Product Spaces

The geometric concepts of distance, angle, and orthogonal projections are important in many applications including signal processing, computer graphics, and approximation of functions. To define these ideas, one needs to introduce inner products on linear spaces.

Inner Products

To extend our study of linear spaces to include geometric entities such as angle between two vectors, distance, shortest distance between a point and a plane, norm of a vector, orthogonality, and other related notions, we shall study *inner products*. Inner products are also referred to as *scalar* or *dot* products.

Initialize the packages

> with(linalg): with(linpdt);

ALGEBRAIC DEFINITION

Let us start with examples from the Euclidean space $\mathbb{R}^n$.

EXAMPLE 1.1   Let u and v be the two vectors in $\mathbb{R}^3$

> u:=vector([4,1,7]); v:=vector([1,3,5]);

Consider the sum of the product of the components of the two vectors

> u[1]*v[1] + u[2]*v[2] + u[3]*v[3];

This product yields a number, not a vector. It can also be evaluated using the Maple command

> w1:=innerprod(u,v);

Consider any two vectors in $\mathbb{R}^3$

> u:=vector([a,b,c]);v:=vector([d,e,f]);

and the sum of the product of the components of the two vectors

> u[1]*v[1] + u[2]*v[2] + u[3]*v[3];

The same quantity can be evaluated using

> 'innerprod(u,v)'=innerprod(u,v);

The product yields a number which is the sum of the product of the corresponding components of the vectors v and u.

The standard **inner product** of any two vectors u and v in the Euclidean space $\mathbb{R}^n$ is a number that is equal to the sum of the product of the corresponding components of the given vectors. Thus, if $u = [a_1, a_2, a_3, \ldots, a_n]$ and $v = [b_1, b_2, b_3, \ldots, b_n]$, then the inner product of u and v denoted by innerprod(u, v) is

```
> S :=Sum(ai*bi,i=1..n);
```

PROPERTIES

What properties must be satisfied by this product to be called an inner product?

Consider the vectors in Example 1.1

```
> u:=vector([4,1,7]); v:=vector([1,3,5]);
```

Compute innerprod(u, v) and innerprod(v, u)

```
> 'innerprod(u,v)'=innerprod(u,v);
  'innerprod(v,u)'=innerprod(v,u);
```

The inner product of two vectors u and v is commutative; that is, innerprod(u,v) = innerprod(v,u). (Fact 4.1)

EXAMPLE 1.2   Consider a vector in $\mathbb{R}^2$

```
> v:=vector([x,y]);
```

What if we take the inner product of v with itself?

```
> innerprod(v,v);
```

This number is always positive. The square root of this number represents the distance from the origin to the point [x, y] in $\mathbb{R}^2$.
   Consider a vector in $\mathbb{R}^3$

```
> v:=vector([x,y,z]);
```

What if we take the inner product of v with itself?

```
> innerprod(v,v);
```

This number is always positive. The square root of this number represents the distance from the origin to the point [x, y, z] in $\mathbb{R}^3$.

The expressions in Example 1.2

1. represent, geometrically, the length (magnitude) of the vectors and
2. indicate that the inner product of a vector with itself is always positive unless the vector is the zero vector in which case the product is 0 (Fact 4.1).

How does the multiplication of a vector by a scalar affect the inner product?

EXAMPLE 1.3   Consider the two vectors

```
> v:=vector([x,y,z]);u:=vector([a,b,c]);
```

Consider the following products

```
> [lambda*u,v]=innerprod(lambda*u,v);
> lambda*[u,v]=lambda*innerprod(u,v);
```

Compare the two numbers. Are they equal? This example shows that

$$\mathbf{innerprod(k * v, u) = k * innerprod(v, u) \ (Fact4.1).}$$

How does this product affect the addition of vectors?

EXAMPLE 1.4   Consider the vectors

```
> v:=vector([x,y,z]); u:=vector([a,b,c]);
w:=vector([p,q,r]);
```

Compute the sum of the following products

```
> '[v,u]+[v,w]'= innerprod(v,u)+innerprod(v,w);
```

and compare this to product:

```
> '[v,u+w]'=innerprod(v,u+w);
```

This example shows that

$$\mathbf{innerprod(v, u + w) = innerprod(v, u) + innerprod(v, w \ (Fact \ 4.1).}$$

SUMMARY

A product defined on a linear space V is called an inner product if the following properties hold for any vectors u, v, and w in V and any real scalars k_1 and k_2:

- The inner product of two vectors is commutative; that is, $< u, v > = < v, u >$
- The innerproduct of a vector with itself is always positive unless the vector is a zero vector; that is, $< u, u > \geq 0$ with equality when u is the zero vector.
- The inner product is linear in the sense that the inner product of $k_1 * u + k_2 * v$ and w is equal to the sum of the inner products of $k_1 * u$ with w and $k_2 * v$ with w; that is,
  $< k_1 * u + k_2 * v, w > = k_1 * < u, w > + k_2 * < v, w >$
  where innerprod(v,u) is denoted by $[v, u], < v, u >$ or $v \cdot u$. All symbols will be used to represent an inner product.

Geometric Definition: Length and Angle Between Vectors

The length of a vector is $L = \sqrt{<v,v>}$ A **unit vector** is a vector whose length is one.

EXAMPLE 1.5   Consider a vector in $\mathbb{R}^3$

```
> v:=vector([x,y]);
```

The length of the vector is

```
> L:=sqrt(innerprod(v,v));
```

Consider another vector in $\mathbb{R}^3$

```
> v:=vector([x,y,z]);
> L:=sqrt(innerprod(v,v));
```

Is there any relation between the angle between two vectors and their inner product? Let us recall from **trigonometry** the cosine law

```
> a^2=b^2+c^2-2*b*c*cos(A);
```

where a, b, and c are the lengths of the sides BC, AC, and AB of a triangle ABC.

EXAMPLE 1.6   Consider the vectors in the plane

```
> oA:=vector([1,1]); oB:= vector([1,-1]);
oC:=vector([0,2]);
```

Construct the sides of the triangle ABC as

```
> AB:=vector([0,-2]); AC:=vector([-1,1]);
BC:=vector([-1,3]);
```

Compute the lengths of the three vectors

```
> a:=sqrt(innerprod(BC,BC)); b:=sqrt(innerprod(AC,AC));
c:=sqrt(innerprod(AB,AB));
```

The cosine law implies that

```
> cos(A):=(b^2+c^2-a^2)/(2*b*c);
```

Compute the ratio

```
> innerprod(AB,BC)/(b*c);
```

This calculation suggests that the **inner product** can be defined

$$< u, v >= innerprod(u, v) = \|v\| * \|u\| * \cos(x)$$

where x is the angle between the two vectors u and v and $\|.\|$ represents the length of the vector.

From the geometric definition of the inner product, we make two inferences

- Two nonzero vectors u and v are orthogonal if and only if innerprod(u, v) =< u, v >= 0.
- | < u,v > | ≤ ‖u‖ * ‖v‖and the equality holds when one vector is a scalar multiple of the other. (Cauchy-Schwartz inequality, Fact 4.2)

Let us use the geometric definition of inner product to find the length of a vector, the angle between two vectors and construct a unit orthogonal vector to given vectors.

EXAMPLE 1.7   Consider two vectors in $\mathbb{R}^3$

```
> v1:=vector([1,-1,0]); v2:=vector([0,1,0]);
```

Determine the length of each vector

```
> L1:=sqrt(innerprod(v1,v1));
L2:=sqrt(innerprod(v2,v2));
```

Determine the inner product of v_1 and v_2

```
> [v1,v2]=innerprod(v1,v2);
```

Determine the cosine of the angle between the two vectors

```
> cos(alpha):=innerprod(v1,v2)/(L1*L2);
```

Determine a unit vector orthogonal to both vectors v_1 and v_2. Let w be the unknown unit vector:

```
> w:=vector([x1,x2,x3]);
```

The following equations must be satisfied for the unit vector w to be orthogonal to v_1 and v_2.
Since vector w is a unit vector

```
> eq1:= innerprod(w,w)=1;
```

Since vector w is orthogonal to the vectors v_1 and v_2

```
> eq2:=innerprod(w,v1)=0;
> eq3:=innerprod(w,v2)=0;
```

Solve the equations for x_1, x_2, and x_3

```
> solve({eq1,eq2,eq3},{x1,x2,x3});
```

How many unit vectors w are orthogonal to v_1 and v_2?

APPLICATIONS

• Closest Point to a Vector

EXAMPLE 1.8   Draw the vector v from $(0,0)$ to the point $(3,1)$. Find a point on v that is closest to the point $(1,2)$.

Let (x,y) be any point on vector v. Since point $P(x,y)$ is on vector v, the point must satisfy $x = 3y$. Vector v and point (x,y) are then given

```
> v:=vector([3,1]);P:=vector([3*y,y]);
```

To find the point P closest to the point $(1,2)$, construct the vector joining the point $P(x,y)$ and the point $(1,2)$

```
> w:=vector([3*y-1,y-2]);
```

such that it is orthogonal to vector P, that is,

```
> innerprod(w,P)=0;
```

The required point is

```
> P:=[3/2,1/2];
```

• Equation of a Plane

EXAMPLE 1.9   Find the equation of the plane passing through the point $Q(1,3,5)$ and normal to the vector $N = [1,-1,4]$. (The **normal** to a plane is the vector orthogonal to every vector in the plane.) To find the equation of the plane, take any point P in the plane

```
> P:=[x,y,z];
```

and construct the vector joining P and Q

```
> PQ:=vector([x-1,y-3,z-5]);
```

The normal to the plane is

```
> N:=vector([1,-1,4]);
```

Since the vector N is orthogonal to every vector in the plane, the equation of the plane is

```
> innerprod(PQ,N)=0;
```

EXERCISES

In the following exercises you may need to use the Maple commands `innerprod`, `vector`, `matrix`, `solve`.

1. a. Find a unit vector

```
> w:=vector([x1,x2,x3,x4]);
```

orthogonal to the vectors

```
> v1:=vector([0,1,1,0]); v2:=vector([1,0,1,0]);
```

b. How many such unit vectors can you find?

2. Determine the equation of the plane passing through the three points P(1, 0, 2), Q(−1, 1, 1), and R(1, 0, 0).

3. Consider the two vectors

```
> v1:=vector([1,-1,0]); v2:=vector([0,1,-1]);
```

a. Find the cosine of the angle between the two vectors.
b. Compare and compute innerprod(v_1, v_2), innerprod(v_1, v_1), innerprod(v_2, v_2), innerprod($v_1 + v_2, v_1 + v_2$).

4. Consider the vector

```
> v:=vector([a,b,c]);
```

Describe the set of all vectors

```
> w:=vector([x,y,z]);
```

that are orthogonal to v.

5. Let u and v be two vectors in $\mathbb{R}^3$. Show that the following inequality holds:

$$\|u + v\| \leq \|u\| + \|v\| \text{(triangle inequality, Fact 4.3)}$$

6. Let $u = [u_1, u_2, u_3]$ and $v = [v_1, v_2, v_3]$ be any two vectors in $\mathbb{R}^3$. Consider the following products in $\mathbb{R}^3$:
$< u, v >= u_1 v_1 + u_2 v_2 - u_3 v_1$
$< u, v >= u_1 v_1 + 2u_2 v_2 + 3u_3 v_3$
Which represents an inner product? Explain.

Orthogonal Projections

The notion of orthogonal projection arises in many applications. For example, in physics when studying the motion of a moving particle along an inclined plane, we analyze the motion by decomposing the force acting on the particle into a component along the direction, u, of the moving particle and a component orthogonal to direction u. In the absence of a solution to a system of linear equations $Ax = b$, one may be interested in constructing an approximate solution to the system. The notion of orthogonal projection plays an important role in this construction.

Initialize the packages

```
> with(linalg):with(linpdt);
```

WHAT IS AN ORTHOGONAL PROJECTION?

Let us begin with the following problem. Given two vectors v and u. How do we construct a vector p along v that is orthogonal to $w = u - p$?

EXAMPLE 2.1   Consider the two vectors

```
> v:=vector([x,y]); u:=vector([l,m]);
```

The vector p is the "orthogonal projection" of u upon v. Since vector p is along the same line as vector v, it follows that p is a multiple of v; that is, $p = t * v$ where t is a scalar to be determined subject to the condition that p is orthogonal to the vector $u - p$; that is, innerprod$(p, u - p) = $ innerprod$(t * v, u - t * v) = 0$. This is equivalent to

```
> t:=innerprod(v,u)/innerprod(v,v);
```

Vector p is then given by

```
> p:=evalm(t*v);
```

and vector w is

```
> w:=evalm(u-p);
```

Is w orthogonal to p? Compute the inner product of w and p

```
> innerprod(p,w);
```

Vector p is called the **orthogonal projection** of u on v, t is the **scalar projection**, and w is the **orthogonal complement**.

EXAMPLE 2.2   Consider the vectors

> `> v:=vector([7,6]); u:=vector([4,2]);`

Compute the orthogonal projection of u upon v. First compute the scalar projection of u on v

> `> t:=innerprod(u,v)/innerprod(v,v);`

Therefore, the orthogonal projection p of u on v is

> `> p:=evalm(t*v);`

The component of u orthogonal to v is

> `> w:=evalm(u-p);`

Vector p is the orthogonal projection of u on v and t is the scalar projection.

   In general, if u and v are two vectors, then

The ratio t = innerprod(u,v)/innerprod(v,v) is called the scalar projection of u on v. The vector p = t ∗ v is the orthogonal projection of u on v. The vector w = u − p is referred to as the orthogonal complement of the vector v. Vector w = u − p is the component of u orthogonal to v. (Fact 4.4)

ORTHOGONAL COMPLEMENT

The orthogonal complement W of a set V is the set of all vectors w orthogonal to every vector in the span of V.

EXAMPLE 2.3   Let us consider set V spanned by the vectors

> `> v1:=vector([1,0,0]); v2:=vector([0,1,0]);`

Describe **analytically** the orthogonal complement of set V. This is the set of all vectors

> `> w:=vector([x,y,z]);`

orthogonal to v_1 and v_2. To determine this set, we compute

> `> innerprod(w,v1)=0; innerprod(w,v2)=0;`

Therefore, the orthogonal complement is W = {[0,0,z] | z ∈ ℝ}. **Geometrically**, this set is the z-axis.

EXAMPLE 2.4   Compute the orthogonal complement of the row space and the column space of the 2 × 2 matrix

```
> A:=matrix([[1,-1],[1,0]]);
```

1. **Orthogonal complement of the row space,** The row space of a matrix A is spanned by

```
> R:=rowspace(A);
```

Let w be a vector in the orthogonal complement:

```
> w:=vector([x,y]);
```

Then:

```
> innerprod(R[1],w)=0; innerprod(R[2],w)=0;
```

Therefore the orthogonal complement of the row space is the set consisting of $\{(0,0)\}$. On the other hand, since A is a nonsingular matrix, the nullspace of the matrix A — that is, the set of all vectors x such that $Ax = 0$ consists of $x = [0,0]$. In this example, the nullspace(A) is the same as the **orthogonal complement of the row space** of matrix A.

2. **Orthogonal complement of the column space**: The column space of matrix A is spanned by

```
> C:=colspace(A);
```

Let w be a vector in the orthogonal complement:

```
> w:=vector([x,y]);
```

Then:

```
> innerprod(C[1],w)=0; innerprod(C[2],w)=0;
```

Therefore, the orthogonal complement of the column space is the set consisting of $\{[0,0]\}$. On the other hand, since A is nonsingular, the nullspace of A^T consists of the trivial set $\{[0,0]\}$. In this example, nullspace(A^T) is the same as the **orthogonal complement of the column space** of the matrix A.

What is the relation between the orthogonal complement of the row space of a matrix and its nullspace?

What is the relation between the orthogonal complement of the column space of a matrix and the nullspace of its transpose?

EXAMPLE 2.5 Compute the orthogonal complement of the row space and the column space of the 2×3 matrix

```
> A:=matrix([[1,-1,1],[1,1,0]]);
```

1. **Orthogonal complement of the row space**: The row space of matrix A is spanned by

```
> R:=rowspace(A);
```

Let w be a vector in the orthogonal complement:

```
> w:=vector([x,y,z]);
```

Then:

```
> eq1:=innerprod(R[1],w)=0;eq2:=innerprod(R[2], w)=0;
> solve({eq1,eq2},{x,y,z});
```

Therefore, the orthogonal complement of the row space is the set consisting of the one-dimensional space with basis $\{[-1, 1, 2]\}$. On the other hand, if we compute the null space of the matrix A:

```
> nullspace(A);
```

In this example, nullspace(A) is the same as the orthogonal complement of the row space of the matrix A

2. Orthogonal complement of the column space

```
> A:=matrix([[1,-1,1],[1,1,0]]);
```

The column space of matrix A is spanned by

```
> C:=colspace(A);
```

Let w be a vector in the orthogonal complement

```
> w:=vector([x,y]);
```

Then

```
> innerprod(C[1],w)=0; innerprod(C[2],w)=0;
```

Therefore, the orthogonal complement of the column space is the set consisting of $\{[0, 0]\}$. Verify that the null space of A^T is also the trivial set. In this example, we find also that nullspace(A^T) is the same as the orthogonal complement of the column space of matrix A.

In general the following relations hold for a given nxm matrix A:

- The nullspace(A) is the same as the **orthogonal complement of the row space** of the matrix A.
- The nullspace(A^T) is same as the **orthogonal complement of the column space** of the matrix A.

We now define **Inner Product** and **Normed Linear Spacse**.

INNER PRODUCT SPACE

An inner product space V is a linear space in which a product $< v, u >$ is defined and satisfies the following properties for all v, u, and w in V and for any real scalars k_1 and k_2:

- $< v, v > \geq 0$ with equality if and only if $v = 0$

- $< v, u > = < u, v >$
- $< k_1 * u + k_2 * v, w > = k_1 * < u, w > + k_2 * < v, w >$

$< v, u >$ with these properties is called an inner product. If V is an inner product space, then the norm of v in V is $\|v\| = \sqrt{< v, v >}$

NORMED LINEAR SPACE

A linear space V is a normed linear space if the following conditions are met

- $\|v\| \geq 0$ with equality if and only if $v = 0$ for any vector v in V.
- $\|av\| = |a| \|v\|$ for any scalar a and any vector v in V.
- $\|u + v\| \leq \|u\| + \|v\|$ for any u and v in V, and equality holds if one vector is a multiple of the other. This property is called the triangle inequality.

All the examples considered so far involve vectors from the Euclidean spaces $\mathbb{R}^n$. There is no reason to restrict ourselves to such spaces. We can choose, for example, the space $C[-1, 1]$ of all continuous functions defined over the interval $[-1, 1]$ or the space $C[-\pi, \pi]$ of all continuous periodic functions over the interval $[-\pi, \pi]$. For example, on $C[-1, 1]$ we can define the product

```
> '< f,g>' = Int (f(x) * g(x), x=-1..1);
```

Is this product an inner product? Check

```
> '< f,f >' =Int(f(x) * f(x), x=-1..1);
```

which is always positive unless $f = 0$.

```
> '< f,g > '=Int(f(x) * g(x), x= -1..1);
'< g,f >' = Int(f(x) * g(x), x=-1..1);
```

which shows that the product is commutative.

```
> '<k1*f+k2*g,h>'= Int((k1*f(x)+k2*g(x))*h(x),x= -1..1);
> 'k1<f,h> +k2<g,h>'= Int(k1*f(x)*h(x),x=-1..1)
  + Int(k2*g(x) *h(x),x=-1..1);
```

Thus the product is linear. Therefore, this is an inner product and $C[-1, 1]$ equipped with this product is an inner product space.

LEAST SQUARES APPROXIMATION

Least squares problems were introduced in Application 3.1. System $Ax = b$ does not have a solution if b is not in the column space of A. In this case an approximate solution is sought. This problem can be stated:

Find an x that minimizes the norm $\|b - Ax\|$.

Algorithm: If the vector b is in the column space of A, then x is the solution to $Ax = b$ and we are done. If b is not in the column space of A, then find the projection b_1 of b upon the column space of A. Now $b - b_1$ is orthogonal to the column space of A; that is, $b - Ax_1$ is orthogonal to each column of A where $Ax_1 = b_1$. From this we may deduce that

$$A^T Ax = A^T b$$

This is called the **normal equation**. The solution x of the normal equation is the least squares solution to the original system.

EXAMPLE 2.6   Given the matrices

```
> A:=matrix([[4,0],[0,2],[2,2]]);
  b:=matrix([[2],[0],[5]]); x:=matrix([[x1],[x2]]);
```

find the least square solution to

```
> evalm(A)*evalm(x)=evalm(b);
```

The system $Ax = b$ is inconsistent. We can verify this by writing the augmented matrix

```
> AUG:=matrix([[4,0,2],[0,2,0],[2,2,5]]);
```

and applying Gauss elimination

```
> ReducedMatrix:=gausselim(AUG);
```

The reduced augmented matrix implies that the system is inconsistent since b is not in the column space of A. Therefore we seek an approximate solution.

LEAST SQUARES PROCEDURE

Premultiply the original system $Ax = b$ by the A^T to get the new system

$$A^T Ax = A^T b$$

Construct the matrix $A^T A$ by premultiplying A by its transpose

```
> trAA:=multiply (transpose(A),A);
```

Construct the matrix $A^T b$ by premultiplying b by the A^T

```
> trAb:=multiply(transpose(A),b);
```

The new system $A^T Ax = A^T b$ is consistent.

```
> evalm(trAA)*evalm(x) = evalm(trAb);
```

Construct the augmented matrix of the new system:

```
> AUG1:=matrix([[20,4,18],[4,8,10]]);
```

Apply Gauss elimination

```
> AUG2:=gausselim(AUG1);
```

Apply backsubstitution to the reduced matrix AUG2 to get the least squares solution

```
> x1 := backsub(AUG2);
```

Compute the error $\|Ax_1 - b\|$.

LEARNING THE PROCESS

Use the automated function `leastsqrs` to obtain the least squares solution. We redo Example 2.6 with

```
> AUG:=matrix([[4,0,2],[0,2,0],[2,2,5]]);
> leastsqrs(AUG);
```

Repeat as necessary with your own matrices until you have mastered the process.

LEAST SQUARES GRAPHICAL DEMONSTRATION

The purpose of this demonstration is to find a polynomial of desired degree that best fits a list of points. You can enter any list of points and the required degree of the least square polynomial. For example, find a polynomial of degree 3 that fits the list of points

```
> L:={[1,2],[2,1],[3,3],[3,-1],[4,2],[-2,2]}:
> lsqrdemo();
```

EXERCISES

In the following exercises, you may need to use `leastsqrs`, and the Maple commands `innerprod`, `crossprod`, `solve`, `vector`, and `matrix`.

1. Compute the orthogonal projection of the vector v on the vector u

```
> v:=vector([1,2,3,-1]); u :=vector([-1,0,5,7]);
```

2. Compute the orthogonal projection of

```
> w :=vector([6,3,-2]);
```

onto the subspace spanned by the vectors

```
> v:=vector([3,4,1]); u :=vector([-2,1,1]);
```

3. Find the closest point to the vector w in the subspace S spanned by the vectors v_1 and v_2

```
> w:=vector([2,1,5,1]); v1:=vector([2,1,-1,1]);
```

```
v2:=vector([0,-1,1,-1]);
```

4. Find the least squares approximation to the system Ax = b, where

```
> A:=matrix([[1,1,0],[1,1,0],[1,0,1],[1,0,1],[1,0,0]]);
  b:=matrix([[-1],[2],[3],[2],[4]]);
  x:=matrix([[x1],[x2],[x3]]);
```

5. Find the least squares approximation to the system Ax = b where

```
> A:=matrix([[1,1,0,0],[1,1,0,0],[1,0,1,0],
  [1,0,1,0],[1,0,0,1],[1,0,0,1]]);
  b:=matrix([[-1],[2],[3],[2],[4],[5]]);
  x:=matrix([[x1],[x2],[x3],[x4]]);
```

Is there a unique least squares solution? Interpret your answer.

6. Find the orthogonal complement of the row space and column space of the matrix

```
> A:=matrix([[1,-1,2,3,1],[2,0,1,-2,4],[3,9,1,- 4,2]]);
```

7. Consider the standard integral inner product on the space C[−1, 1]. That is, for any two functions f and g in C[−1, 1], we define the inner product of f and g by

```
> fg:=Int(f(x)*g(x),x=-1..1);
```

Let

$$B = \left\{ \frac{1}{\sqrt{2}}, \sqrt{\frac{3}{2}}x \right\}$$

be a basis for the set of all polynomials of degree less or equal to 1.

a. Compute the orthogonal projection of the function e^{2x} on the subspace spanned by the set B.

b. Deduce the least squares linear approximation $a + bx$ for the function e^{2x}.

Gram-Schmidt Orthogonalization

A basis of a vector space is its building block. Every element of the vector space can be uniquely expressed in terms of the basis. In the Euclidean space $\mathbb{R}^n$, it is easy to express a vector in terms of the standard basis. However, the process can be very tedious if the basis is not the standard one. Another special and useful basis is the **orthonormal basis**. Every vector can be easily expressed in terms of the orthonormal basis.

Initialize the packages

```
> with(linalg):with(linpdt);
```

ORTHOGONAL SETS

EXAMPLE 3.1   Let $B = \{u, v\}$ be a basis for $\mathbb{R}^2$

```
> u:=vector([-2,3]); v:=vector([3,2]);
```

Let w be any vector in $\mathbb{R}^2$

```
> w:=vector([x,y]);
```

Express the vector w in terms of B:

```
> evalm(w)=c1*evalm(u) +c2*evalm(v);
```

The problem of expressing w in terms of the basis reduces to solving the following system for c_1 and c_2:

```
> eq1:= -2*c1 +3*c2 = x; eq2:= 3*c1 +2*c2 = y;
> solve({eq1,eq2},{c1,c2});
```

Therefore,

```
> c1:=3/13*y-2/13*x; c2:=2/13*y+3/13*x;
> evalm(w)=c1*evalm(v)+c2*evalm(u);
```

Let us look closely at set B. Form the inner product of the vectors u and v

```
> innerprod(u,v);
```

The two vectors u and v are orthogonal. Instead of solving a system of equations for c_1 and c_2 to represent the vector w in terms of the basis vectors u and v, let us see if we can obtain the values of c_1 and c_2 differently.

If $w = c_1 * u + c_2 * v$, then the inner product of w with u and with v respectively yields

```
> c1:=innerprod(w,u)/innerprod(u,u);
> c2:=innerprod(w,v)/innerprod(v,v);
```

Compare with the values obtained earlier for c_1 and c_2. Thus, if the basis is orthogonal, then it is much easier to solve for c_1 and c_2. Set B is called an **orthogonal basis** for $\mathbb{R}^2$.

EXAMPLE 3.2   Consider a basis B in $\mathbb{R}^3$ consisting of the vectors

```
> u1:=vector([-1,1,1]); u2:=vector([2,1,1]);
u3:=vector([0,-5,5]);
```

Is this an orthogonal basis? Form the inner products

```
> innerprod(u1,u2); innerprod(u1,u3); innerprod(u2,u3);
```

Thus, set B is an orthogonal basis. Try to express any vector w in $\mathbb{R}^3$

```
> w:=vector([x,y,z]);
```

as a linear combination ($w = c_1u_1 + c_2u_2 + c_3u_3$) of basis B. Since the vectors are mutually orthogonal, the inner product of w with u_1, u_2 and u_3 will give

```
> c1:=innerprod(w,u1)/innerprod(u1,u1);
> c2:= innerprod(w,u2)/innerprod(u2,u2);
> c3:=innerprod(w,u3)/innerprod(u3,u3);
```

For example, if w is the vector

```
> w:=vector([-1,2,3]);
```

then it can be expressed in terms of u_1, u_2, and u_3 with the following values of c_1, c_2 and c_3:

```
> c1:=innerprod(w,u1)/innerprod(u1,u1);
> c2:= innerprod(w,u2)/innerprod(u2,u2);
> c3:=innerprod(w,u3)/innerprod(u3,u3);
```

Thus, vector $w = [-1, 2, 3]$ is equal to

```
> evalm(w)=c1*evalm(u1)+ c2*evalm(u2) + c3*evalm(u3);
```

A set of vectors $\{v_1, v_2, \ldots, v_n\}$ in an inner product space V is said to be **orthogonal** if and only if the vectors are mutually orthogonal; that is, $innerprod(v_i, v_j) = 0$ for all $i, j (i \neq j)$.

If $B = \{u_1, u_2, \ldots, u_n\}$ is an orthogonal basis of an inner product space V, then any vector w in V can be written as a linear combination:

$$w = c_1u_1 + c_2u_2 + c_3u_3 + \cdots + c_nu_n$$

where $c_i = \frac{\text{innerprod}(w, u_i)}{\text{innerprod}(u_i, u_i)}$ for $i = 1, 2, \ldots, n$. (Fact 4.5)

ORTHOGONAL SETS AND LINEAR INDEPENDENCE

Is every orthogonal set of vectors linearly independent? Is the converse true?

EXAMPLE 3.3   Let us consider the following set in $\mathbb{R}^4$

```
> u1:=vector([1,1,-1,0]); u2:=vector([0,0,0,3]);
u3:=vector([0,-1,-1,0]);
```

Are these vectors mutually orthogonal?

```
> innerprod(u1,u2); innerprod(u1,u3); innerprod(u2,u3);
```

Do they form an independent set of $\mathbb{R}^4$? A set of vectors is linearly independent if a zero combination of the set $\{u_1, u_2, u_3\}$, that is, $c_1 u_1 + c_2 u_2 + c_3 u_3 = 0$, implies that all the scalars are zero. Since the vectors u_1, u_2, and u_3 are orthogonal, the inner product of these vectors with the linear combination implies $c_1 = c_2 = c_3 = 0$. Therefore, the vectors are independent.

EXAMPLE 3.4   Consider an independent set of vectors in $\mathbb{R}^3$

```
> u1:=vector([1,1,-1]);u2:=vector([0,0,3]);
> u3:=vector([0,-1,-1]);
```

Do they form an orthogonal basis for $\mathbb{R}^3$?

If $B = \{v_1, v_2, \ldots, v_n\}$ is a non-empty subset of mutually orthogonal nonzero vectors in an inner product space then set B is linearly independent. (Fact 4.6)

Is the converse of this statement true? That is, if the set of vectors $B = \{v_1, v_2, \ldots, v_n\}$ is linearly independent, is it true that set B is an orthogonal set?

ORTHONORMAL BASIS (ONB)

EXAMPLE 3.5   Consider the orthogonal set of vectors

```
> u1:=vector([1,1,-4]); u2:=vector([2,2,1]);
u3:=vector([1,-1,0]);
```

The length of each vector is

```
> Lu1:=sqrt(innerprod(u1,u1));
Lu2:=sqrt(innerprod(u2,u2));
Lu3:=sqrt(innerprod(u3,u3));
```

Dividing each vector by its length yields three mutually orthogonal unit vectors

```
> w1:=evalm(u1/Lu1); w2:=evalm(u2/Lu2);
w3:=evalm(u3/Lu3);
```

Does this set form a basis for $\mathbb{R}^3$? This set is linearly independent. Hence, it is a basis for $\mathbb{R}^3$. A basis consisting of mutually orthogonal unit vectors is called an **orthonormal basis**.

Consider the vector

```
> w:=vector([-5,56,72]);
```

and express it as a linear combination of the vectors w_1, w_2, and w_3; that is, $w = c_1 w_1 + c_2 w_2 + c_3 w_3$.

```
> c1:=innerprod(w,w1); c2:=innerprod(w,w2);
c3:=innerprod(w,w3);
```

Therefore,

```
> w:=c1*w1+ c2*w2 +c3*w3;
```

and verify that this is exactly w,

```
> evalm(w)=c1*evalm(w1)+ c2*evalm(w2) + c3*evalm(w3);
```

An orthonormal basis $v_1, v_2, \ldots, v_n$ of an inner product space V is a basis whose elements are mutually orthogonal unit vectors.

If $\{v_1, v_2, \ldots, v_n\}$ is an orthonormal basis for an inner product space V and if w is any vector in V, then

$$w = c_1 * v_1 + c_2 * v_2 + c_3 * v_3 + \cdots + c_n * v_n,$$

where $c_i = \text{innerprod}(w, v_i)$.

GRAM-SCHMIDT CONSTRUCTION OF AN ONB

Check whether the given set is linearly independent. If it is independent, proceed to construct an orthonormal basis; otherwise, extract an independent set and then proceed. Assume that the set $B = \{v_1, v_2, \cdots, v_n\}$ is independent.

1. construct the first vector

$$u_1 = v_1 / \|v_1\|$$

2. Find the orthogonal projection of v_2 on u_1

$$p_1 = \text{innerprod}(v_2, u_1)u_1$$

and construct the vector $w_2 = v_2 - p_1$. Then the second member of the orthonormal basis is $u_2 = w_2 / \|w_2\|$.

3. Find the projection of v_3 onto $\text{span}(u_1, u_2)$

$$p_2 = \text{innerprod}(v_3, u_2)u_2 + \text{innerprod}(v_3, u_1)u_1$$

and construct the vector $w_3 = v_3 - p_2$. The third element in the orthonormal basis is

$$u_3 = w_3/\|w_3\|$$

4. Repeat the steps until all vectors of the original set have been exhausted. The resulting set $\{u_1, u_2, \cdots, u_n\}$ is an orthonormal basis. This process is call the Gram-Schmidt Orthogonalization. **(Fact 4.7)**

LEARNING THE PROCESS

To learn the Gram-Schmidt process, execute the demonstration mode of the function `GramSchmidt`. Here is an example:

```
> v1:=vector([3,4]); v2:=vector([6,8]);
v3:=vector([-1,1]);
> GramSchmidt(v1,v2,v3);
```

Work several of your own examples before you proceed.

ORTHOGONAL MATRICES

An $n \times n$ matrix is said to be an **orthogonal matrix** if its columns form an orthonormal set.

EXAMPLE 3.7   Consider the orthonormal basis of $\mathbb{R}^3$

```
> u1:=vector([3/5,4/5,0]); u2:=vector([-4/5,3/5,0]);
u3:=vector([0,0,1]);
```

Construct the matrix whose columns are the vectors u_1, u_2, and u_3

```
> A:=augment(u1,u2,u3);
```

The columns of this matrix form an **orthonormal set**. Thus the matrix A is called an **orthogonal matrix**. Let us study some of its properties. Take, for example, the A^T:

```
> trA:=transpose(A);
```

What is the product of A and A^T ?

```
> AtrA=multiply(A,trA);
```

What is the determinant of the matrix A? What is the inverse of the orthogonal matrix A? In general, an orthogonal matrix A has the following properties

- Its inverse is equal to its transpose; that is $A^{-1} = A^T$.

- Its determinant is equal to $+1$ or -1.

EXERCISES

In the following exercises you may need to use the function GramSchmidt and the Maple commands innerprod, crossprod, solve, vector, matrix, and evalm.

1. Find an orthogonal basis for the column space of the matrix

```
> A:=matrix([[1,3,5],[-1,-3,1],[0,1,3],[1,5,3]] );
```

2. Construct an orthonormal basis, using the interactive mode of the GramSchmidt function, for the subspace spanned by the vectors

```
> v1:=vector([1,-1,1,1]); v2:=vector([0,1,2,6]);
    v3:=vector([0,1,0,1]); v4:=vector([1,1,3,1]);
```

Express the vector

```
> w:=vector([1,-1,6,7]);
```

in terms of the orthonormal basis.

3. a. Construct an orthonormal basis for the set of all polynomials of degree less than or equal to 2 from the set of independent functions $\{1, x, x^2\}$

```
> f:=x->1; g:=x->x; h:=x->x^2;
```

using the inner product

```
> pq:=Int(p(x)*q(x), x=-1..1);
```

b. Extend the basis constructed in (a) to an orthonormal basis for the set of all polynomials of degree less than or equal to 3.
c. Approximate the function

```
> s:=x->x^3+2*x-1;
```

as a linear combination of the basis constructed in (b).

4. a. Compute values of a, b, and c so that a matrix A is an orthogonal matrix

```
> A:=matrix([[1/sqrt(2),0,1/sqrt(2)],
    [1/sqrt(2),0,-1/sqrt(2)], [a,b,c]]);
```

b. Compute values of a, b, and c so that matrix A is an orthogonal matrix

```
> A:=matrix([[1/sqrt(3),1/sqrt(3),1/sqrt(3)],
    [a,b,c],[1/sqrt(2),0,-1/sqrt(2)]]);
```

c. Compute values of a, b, c, d, e, f, g, and h so that matrix A is an orthogonal matrix

```
> A:=matrix([[1/3,1/3,a,b],[1/3,-1/3,c,d],
    [1/3,1/3,e,f],[1/3,-1/3,g,h]]);
```

5. This exercise is a continuation of Exercise 3 of this lesson and Exercise 6 of Lesson 4.2. Given the standard integral inner product on the space $C[-1, 1]$ — that is, for any two functions f and g in $C[-1, 1]$ — we define the inner product of f and g by

```
> fg:=Int(f(x)*g(x),x=-1..1);
```

Let $B = \{1, x, x^2, x^3\}$ be a basis for the set of all polynomials of degree less than or equal to 3.

a. Convert set B into an orthonormal basis B_1 using the given inner product.

b. Compute the orthogonal projection of function f in $C[-1, 1]$ on the subspace spanned by set B_1.

c. Deduce the least squares linear (a + bx), quadratic(a + bx + cx^2), and cubic approximation (a + bx + cx^2 + dx^3) for the function e^{2x}.

d. Decide which one of the approximants yields a better approximation for function e^{2x}.

QR-Decomposition

If A is an $m \times n$ matrix, as a result of constructing an orthonormal basis for the column space of matrix A, one can get a factorization of matrix A as a product QR where the columns of matrix Q consist of an orthonormal basis and matrix R an upper triangular matrix with positive entries on the main diagonal. This factorization is useful, for example, in finding inverses, in solving systems of linear equations, and in finding eigenvalues.

Initialize the packages

```
> with(linalg):with(linpdt);
```

DESCRIPTION OF THE QR-ALGORITHM

EXAMPLE 4.1   Consider the matrix

```
> A:=matrix([[1,1,1],[1,0,2],[1,1,0]]);
```

whose columns are the independent vectors

```
> v1:=vector([1,1,1]); v2:=vector([1,0,1]);
v3:=vector([1,2,0]);
```

1. Convert the set of vectors $\{v_1, v_2, v_3\}$ into an orthonormal set. Normalize the first vector v_1 to obtain the first orthonormal vector u_1

```
> r11:=sqrt(innerprod(v1,v1));
u1:= evalm(1/r11*v1);
```

Find a vector w_2 that is orthogonal to u_1

```
> r12:=innerprod(v2,u1);
p1:=evalm(r12*u1): w2:=evalm(v2-p1);
```

Normalize the vector w_2 to obtain the second orthonormal vector u_2

```
> r22:=sqrt(innerprod(w2,w2)); u2:=evalm(1/r22*w2);
```

Compute the third orthonormal vector by computing the inner products

```
> r13:=innerprod(v3,u1); r23:=innerprod(v3,u2);
```

```
p2:=evalm(r23*u2 + r13*u1): w3:=evalm(v3-p2);
```

Normalize the vector w_3 to obtain the third orthonormal vector u_3

```
> r33:=sqrt(innerprod(w3,w3)); u3:=evalm(1/r33*w3);
```

2. Construct the matrix whose columns are the orthonormal vectors u_1, u_2, and u_3

```
> Q:=augment(u1,u2,u3);
```

3. Construct an upper triangular matrix R whose entries are the r_{ij} obtained in step 1

```
> R:=matrix([[r11,r12,r13],[0,r22,r23], [0,0,r33]]);
```

Perform the multiplication of matrix Q with matrix R.

```
> multiply(Q,R);
```

which is the matrix A.

LEARNING THE PROCESS

Use the demonstration mode of QRdecomp to learn this process.

EXAMPLE 4.2   Consider

```
> A:=matrix([[1,0,-1],[1,1,0],[1,1,1]]);
```

Find the QR-factorization of A using the function

```
> QRdecomp(A);
```

Choose your own matrices and repeat until you have mastered the process.

EXERCISES

In the following exercises you may need to use the automated functions QRdecomp and GramSchmidt and the Maple commands solve, evalm, matrix, vector, innerprod, and crossprod.

1. Find the QR decomposition of the matrix

```
> A:=matrix([[1,2,5],[-1,1,-4],[-1,4,-3],[1,-4,7]]);
```

2. Consider the matrices

```
> A:=matrix([[1,2,5],[1,1,0],[1,1,2],[1,3,3]]);
    b:=matrix([[2],[5],[6],[7]]);
```

a. Find the least square solution of $Ax = b$.

b. Obtain the QR decomposition of the matrix A.

c. Show that the least squares solution is given by $x_1 = R^{-1}Q^T b$.

Orthogonal Basis and QR-Decomposition

Purpose

The purpose of this lab is to enforce the basic concepts of inner product spaces, including the GramSchmidt orthogonalization process and QR-decomposition.

Automated Linalg functions

In this lab you will use of the automated functions `GramSchmidt`, `QRdecomp`, and `leastsqrs`. To get help with a function, type at the Maple prompt >? function name; for example, >?GramSchmidt;

INSTRUCTIONS

1. To execute a statement move the cursor to the line using the mouse or the keypad and press the enter key.
2. Create Maple input regions as needed to ensure that the output appears in the desired place.
3. Execute the following commands to load the packages

```
> with(linalg):with(linpdt)
```

TASK 1

The purpose of this task is to learn inner product and some of its implications. Let u and v be two vectors in $\mathbb{R}^3$

```
> u:=vector([1,-2,1]); v:=vector([-3,1,4]);
```

Activity 1

Find the inner product of the two vectors u and v.

Activity 2

Find the cosine of the angle between the two vectors u and v.

Activity 3

Find the orthogonal projection of u on v.

Activity 4

Find a unit vector $w = [a, b, c]$ orthogonal to both u and v.

Activity 5

Find the equation of the plane passing through the point $P(1, 1, 1)$ where w is normal to the plane.

Activity 6

What is the shortest distance between the plane and point $Q(0, 1, -1)$?

TASK 2

The purpose of this task is to construct an orthonormal basis using the GramSchmidt method. As a consequence, we deduce the QR-decomposition of a matrix. Let B be a subset of $\mathbb{R}^3$:

```
> u1:=vector([1,-1,1]);u2:=vector([0,1,0]);
  u3:=vector([1,0,1]);u4:=vector([0,1,2]);
```

Activity 1

Is set B linearly independent? If not, extract an independent subset B_1.

Activity 2

Use subset B_1 to construct an orthonormal basis for the subspace spanned by B_1. Use the interactive mode of GramSchmidt.

Activity 3

Construct the QR-decomposition for the matrix A whose columns constitute the given vectors of the set B_1. Determine matrix Q. Deduce matrix R.

TASK 3

The purpose of this task is to relate QR-decomposition to the notion of pseudo-inverse of a matrix A. Consider the matrix

```
> A:=matrix([[2,1],[1,1],[2,3]]);
```

Activity 1

Construct matrix Q in the QR-decomposition of matrix A. Use the interactive mode of QRdecomp.

Activity 2

Verify that the product QR is equal to matrix A.

Activity 3

Compute the product of the matrices R^{-1} and Q^T.

Activity 4

Compute the product of the matrices $(A^T A)^{-1}$ and A^T.

Activity 5

Compare the results of Activities 3 and 4. Compute the product of $R^{-1}Q^T$ and A. *Note*: The matrices: $R^{-1}Q^T$ or $(A^T A)^{-1}A^T$ is called the **pseudo-inverse** of matrix A. (The matrix A has a one sided inverse.)

TASK 4

The purpose of this task is to apply the pseudo-inverse concept of Task 3 to the problem of fitting a curve through a set of points. Consider the set of points

```
> P1:=[1,2]; P2:=[-1,0]; P3:=[1,1];
P4:=[3,5]; P5:=[8,-3]; P6:=[4,3];
```

Activity 1

Suppose we want to fit a quadratic curve $y = ax^2 + bx + c$ through these points. Write the resulting system of linear equations to solve for a, b, and c. Is there a solution to this system? Justify your answer

Activity 2

Write the system in matrix form $Ax = b$, where A is the coefficient matrix, x is the unknown matrix and b is the matrix whose entries are the values of the equations. Enter matrices A, x, and b.

Activity 3

Construct the pseudo-inverse P of matrix A.

Activity 4

Multiply both sides of $Ax = b$ by the pseudo-inverse P to solve the system.

Activity 5

Write the quadratic curve that fits the given points based on the estimated values of a, b, and c.

Activity 6

Compute the difference between the true value and the estimated value for each point. What is the maximum error?

EXTRA LAB PROBLEMS

1. Compare the solution of the problem in Task 4 to the least squares solution of the problem.

2. Consider the upper triangular matrix

```
> A:=matrix([[a,b,c],[0,d,e],[0,0,f]]);
```

Find values of the parameters a, b, c, d, e, and f if the columns of matrix A are orthonormal. Generalize the result to any $n \times n$ matrix.

SUPERVISED LAB 4.2

Orthogonal Basis and Least Squares Applications

Purpose

The purpose of this lab is to apply concepts of inner product spaces including the GramSchmidt orthogonalization process and QR-factorization to obtain the least squares solution to a linear system.

Automated Linalg functions

In this lab you will use the interactive functions `GramSchmidt`, `QRdecomp`, and `leastsqrs`. To get help with a function, type at the Maple prompt sign >?function-name; for example, >?GramSchmidt;

INSTRUCTIONS

1. To execute a statement move the cursor to the line using the mouse or the keypad and press the enter key
2. Create Maple input regions as needed to ensure that the output appears in the desired place.
3. Execute the following commands to load the packages

```
> with(linalg):with(linpdt);
```

TASK 1

The purpose of this task is to use inner products to construct an approximate solution to a system of linear equations. Consider the system of linear equations given by $Ax = b$ where the coefficient matrices are

```
> A:=matrix([[1,2,-1],[2,3,1],[-1,-1,-2],[3,5,0]]);
  b:=matrix([[1],[0],[1],[0]]);
```

Activity 1

Find a basis B for the column space of matrix A. Does vector b lie in the column space of A ? Is system $Ax = b$ consistent ?

Activity 2

Construct an orthonormal basis B_1 for the column space of A.

Activity 3

Construct the vector b_1 that is a linear combination of the basis B_1 with coefficients being the inner products of vector b with the elements of B_1.

Activity 4

Which of the systems $Ax = b$ and $Ax = b_1$ is consistent? If consistent, find all solutions x.

Activity 5

Can you show that for all x

$$\|Ax - b_1\| \leq \|Ax - b\|$$

in the Euclidean norm? When does the equality hold?

TASK 2

The purpose of this task is to apply least squares solutions and QR-factorization to the solution of a linear regression problem. Find the coefficients of the least squares line

```
> y:= beta0 +beta1*x;
```

that best fit the data

```
> P1:=[1,2]; P2:=[3,2]; P3:=[5,7]; P4:=[7,9];
```

Activity 1

What are the equations satisfied by the coefficients if data points lie on line y?

Activity 2

Write the coefficient matrix A, the predicted value b, and the unknown vector x,

Activity 3

Is the system consistent? If not, compute the least squares solution that best fits the data.

Activity 4

Plot the line of best fit and the data points.

EXTRA LAB PROBLEM

Generalize the results of Task 2 to find the coefficients of the least squares curve

```
> y:= beta0 +beta1*x + beta2*x^2;
```

that best fits the data in Task 2. Compare with the plot in Activity 4.

Population Dynamics

Purpose

The purpose of this application is to apply concepts studied in this unit to solve a population dynamics model.

Initialize the package

```
> with(linalg):with(linpdt);
```

DESCRIPTION OF A POPULATION DYNAMICS MODEL

Two species A and B live in a colony and compete for survival. A grows at a rate of 30% of its current size plus a rate 20% of the current size of B. Species B grows at a rate of 40% of its current size minus a rate k of the current size of A. We want to determine the effect of varying k on the population behavior. Let A_i and B_i represent the number of species A and B respectively in time period i. Furthermore, assume that the original population of species A and B is $A_0 = 100$ and $B_0 = 300$.

Activity 1

Write the mathematical equations that describe the size of the populations at any time period i.

Activity 2

Write the equations in matrix form $D_{i+1} = PD_i$ where $D_i = \text{matrix}([[A_i], [B_i]])$ and P is the coefficients matrix.

Activity 3

Compute the first five powers of matrix P.

Activity 4

a. Choose $k = 0.1$. Compute the population of each species for $i = 1, 2, 3, 4, 5$

b. Choose $k = 0.2$. Compute the population of each species for $i = 1, 2, 3, 4, 5$.

Activity 5

From the computations in Activity 4, what can you infer about the size of each population as you vary the rate k? Explain carefully. Do you think the size of each population will continue to grow? Will there be a period of time when there is a shift in the population behavior? How does this relate to the rate k?

Activity 6

Assume the coefficient matrix:

```
> P:=matrix([[a1,a2],[b1,b2]]);
```

Use least squares to determine the values of a_1, a_2, b_1, and b_2 to fit the population data obtained in Activity 4a, summarized in the following table.

i	0	1	2	3	4	5
A_i	100	190	329	438	849	1301
B_i	300	410	555	744	987	1298

How do the values of a_1, a_2, b_1, and b_2 compare with the entries of matrix P of Activity 2?

Orthogonal Expansions and Signal Processing

Purpose

The purpose of this application is to study a class of orthogonal polynomials and relate them to a problem in signal processing.

Initialize the package

```
> with(linalg):with(linpdt);
```

LEGENDRE POLYNOMIALS

Consider the linear space C[−1, 1] of continuous functions over [−1, 1]. The class of Legendre polynomials can be used to approximate functions in C[−1, 1]. We will develop this class of polynomials and show how to use it to appoximate functions in C[−1, 1]. Start with the independent set of functions

```
> S:={1,x,x^2,x^3,x^4};
```

and define the inner products of two functions in C[−1, 1]

```
> pdt:=int(b(x)*c(x),x=-1..1);
```

Activity 1

Construct the first five orthonormal polynomials $p[i], i = 0, 1, 2, 3, 4$ (*Hint*: Use the GramSchmidt orthogonalization process. Do not use the automated function.) The first two such polynomials are

```
> p[0]:=1/sqrt(2)*1;
> w[1]:=x-int(x*p[0],x=-1..1)*p[0];
> p[1]:=1/sqrt(int(w[1]*w[1],x=-1..1))*x;
```

Activity 2

Suppose that the function $f(x) = \sin(x)$

```
> f:=x-> sin(x);
```

is to be approximated by the function $g(x)$ using these orthogonal polynomials, that is,

```
> g(x) := sum(a[i]*p[i],i=0..4);
```

Determine the coefficients $a[i]$ for $i = 0, 1, 2, 3, 4$. For example $a[0]$ is

```
> a[0]:=int(p[0]*g(x),x=-1..1);
```

Plot the graph of $\sin(x)$ and its approximated value over the interval $[-1, 1]$ on the same coordinate axes. Is the approximation a reasonable one?

```
> plot({f(x),g(x)},x=-1..1);
```

Activity 3

Repeat Activity 2 for the function $f(x) = \cos(x)$

Activity 4

If $p[i], i = 0, 1, 2, \ldots$ is the set of all such orthogonal polynomials constructed from the independent set $\{x^i, \ i = 0, 1, 2, \ldots\}$, determine the general formula for obtaining the nth orthogonal polynomial and the coefficients in the expansion

```
> f:=x->sum(a[i]*p[i],i=0..infinity);
```

TRIGONOMETRIC POLYNOMIALS

Consider the Linear space, $C[0, 2\pi]$, of continuous periodic functions of period 2π and the set $S = \{1, \cos(x), \sin(x), \cos(2x), \sin(2x), \ldots\}$. Define the inner product on $C[0, 2\pi]$

```
> pdt:=int(a(x)*b(x),x=0..2*Pi);
```

Activity 1

Show that the set S is an orthogonal set. (*Hint*: Consider the product of $\cos(mx)$ and $\sin(nx)$ for various values of m and n.)

Activity 2

Convert set S into an orthonormal set.

Activity 3

The polynomial

```
> p :=x->sum(a[k]*cos(k*x) +b[k]*sin(k*x),k=0..n);
```

is an example of an orthogonal trigonometric polynomial. Determine a method for computing coefficients a_k and b_k in this expansion.

Activity 4

Suppose that the function

```
> f := x ->x^2;
```

is to be approximated using the polynomial p(x). Determine the coefficients a_k and b_k for this function f(x). Plot the graph of x^2 and its approximated value over the interval $[0, 2\pi]$ on the same coordinate axes. Is the approximation a reasonable one?

SIGNAL PROCESSING AND SAMPLING

In this application, we determine a continuous signal from a sampled data points. Let f be a function whose values f[0], f[1], f[2], ..., f[n − 1] are known at n equally spaced points $t[0] = 0, t[1] = \frac{2\pi}{n}, t[2] = 2\frac{2\pi}{n}, ..., t[n − 1] = (n − 1)\frac{2\pi}{n}$ of the interval $[0, 2\pi]$.
  Define

```
> i:=sqrt(-1):k:='k':
```

Further, assume that function f is a periodic function

```
> f:=t->sum(a[k]*exp(k*I*t),k=0..n-1);
> f(t);
```

Determine the values of the coefficient a_k subject to the condition that f(t) passes through the points (t[k], f[k]), k = 0, 1, 2, ..., n − 1.

Activity 1

Assume that a signal has the sampled data f[0] = 0, f[1] = 1, f[2] = 2, f[3] = 3 at the sampled points $t[0] = 0, t[1] = \frac{\pi}{2}, t[2] = \pi, t[3] = \frac{3\pi}{2}$.

a. By substituting these values in the expression for f, determine a system of linear equations for the unknown coefficients.
b. Write the equivalent matrix representation of this system.
c. Is the coefficient matrix invertible?
d. Solve the system.
e. Write the function f. Plot f and the sampled data. Is the signal reasonably close to the sampled data?

Activity 2

Can you propose a method that handles n sampled points? Describe the procedure.

Linear Transformations

In this unit we study relations between linear spaces. This leads to the study of linear transformations and their properties. Characterizing special subspaces associated with a linear transformation, namely the kernel and range of a transformation, sheds light on the solvability of the system $Ax = b$. Also, the matrix representation of a linear transformation relative to different bases is an important problem in many applications.

Introduction to Linear Transformations

In this lesson we shall study the properties and representations of mappings between linear spaces. We shall also investigate whether the set of such mappings, under appropriate definitions of addition and scalar multiplication, forms a linear space.

Initialize the packages

```
> with(linalg):with(lintran);
```

WHAT IS A LINEAR TRANSFORMATION?

Let us start with some examples of mappings between vector spaces.

EXAMPLE 1.1   Consider the mapping from $\mathbb{R}^2$ into $\mathbb{R}^2$

```
> T :=(x,y)->(y,x);
```

Does T preserve addition? That is, is $T(v + u) = T(v) + T(u)$ for any two vectors v and u in $\mathbb{R}^2$?

Take any two vectors in $\mathbb{R}^2$

```
> v:=vector([a,b]); u:=vector([c,d]);
```

Compute the action of T on $v + u$

```
> 'T(v + u)':=[T(a+c,b+d)];
```

and the sum $T(v) + T(u)$

```
> 'T(v)':=[T(a,b)];'T(u)':=[T(c,d)];
> 'T(v)+T(u)':= evalm(vector([T(a,b)])+vector([T(c,d)]));
```

Is $T(v + u) = T(v) + T(u)$? It is clear from the computation that $T(v + u) = T(v) + T(u)$. Therefore, the mapping T preserves addition.

Does T preserve scalar multiplication? If k is any real scalar, compare the action of T on $k * v$, that is $T(k * v)$, to $k * T(v)$

```
>  'T(kv)':=[T(k*a,k*b)]; 'kT(v)':=evalm(k*[T(a,b)]);
```

Is $T(k*v) = k*T(v)$? It is clear from the computation that $T(k*v) = k*T(v)$. Therefore, the mapping T preserves scalar multiplication.

This example shows that the transformation T preserves addition and scalar multiplication. Such a mapping T is called a **linear transformation**.

EXAMPLE 1.2 Consider the mapping T from $\mathbb{R}^3$ to $\mathbb{R}^2$

```
>  T  := (x,y,z) -> (x-y,y-z);
```

Does T preserve addition? That is, is $T(v + u) = T(v) + T(u)$ for any two vectors v and u in $\mathbb{R}^3$? Take any two vectors in $\mathbb{R}^3$

```
>  v:=vector([a1,a2,a3]); u:=vector([b1,b2,b3]);
```

Compare the action of T on $v + u$ to the sum of $T(v)$ and $T(u)$

```
>  'T(v1+v2)':=[T(a1+b1,a2+b2,a3+b3)];
>  'T(v1)':=[T(a1,a2,a3)]; 'T(v2)':=[T(b1,b2,b3)];
```

Compute

```
>  'T(v)+T(u)':= evalm(vector([T(a1,a2,a3)])+
vector([T(b1,b2,b3)]));
```

Is $T(v + u) = T(v) + T(u)$? It is clear from the computation that $T(v + u) = T(v) + T(u)$. Therefore, the transformation T preserves addition.

Does the transformation T preserve scalar multiplication? For any scalar k, compare the action of T on $k*v$, that is $T(k*v)$ to $k*T(v)$

```
>  T(kv):=[T(k*a1,k*a2,k*a3)];
kT(v):=evalm(k*[T(a1,a2,a3)]);
```

Is $T(k*v) = k*T(v)$? It is clear from the computation that $T(k*v) = k*T(v)$. Therefore, the transformation T preserves scalar multiplication. Thus T is a **linear transformation**.

EXAMPLE 1.3 Does matrix multiplication define a linear transformation? Choose a matrix and an appropriate vector

```
>  A:=matrix([[1,3],[-2,3]]); X:=matrix([[x],[y]]);
```

Define T from $\mathbb{R}^2$ to $\mathbb{R}^2$ in terms of the matrix multiplication of A and X

```
>  T:=X->multiply(A,X);
>  'T(X)' =T(X);
```

Choose any two vectors v and u in $\mathbb{R}^2$:

```
>  v:=matrix([[a],[b]]);u:=matrix([[c],[d]]);
```

Determine the action of T on $v + u$ and compare it to $T(v) + T(u)$

```
> 'T(v) + T(u)':=evalm(T(v)+T(u)); 'T(v + u )':=T(v+u);
```

Does T preserve addition?

Does T preserve scalar multiplication?

```
> 'kT(v)':=evalm(k*T(v)); 'T(kv)':=T(evalm(k*v));
```

T preserves scalar multiplication. Therefore, matrix multiplication is an example of a linear transformation.

Let us consider another type of transformation.

EXAMPLE 1.4 Consider a transformation on $\mathbb{R}^2$

```
> T :=(x,y)->(x-y,y^2);
```

Does T preserve the operations of addition and scalar multiplication? Take any two vectors in $\mathbb{R}^2$

```
> v:=vector([a,b]); u:=vector([c,d]);
```

Is $T(v + u) = T(v) + T(u)$?

```
> 'T(v+u)':=[T(a+c,b+d)];
> 'T(v)+T(u)':= evalm(vector([T(a,b)])+vector([T(c,d)]));
```

Compare $T(v + u)$ and $T(v) + T(u)$. Are they equal?

In this example, the transformation T does not preserve addition. Such a transformation is a nonlinear transformation.

SUMMARY

A linear transformation T is a mapping from a vector space V into a vector space W satisfying two conditions:

- T is linear (T preserves addition):
  for any v_1 and v_2 in V, $T(v_1 + v_2) = T(v_1) + T(v_2)$.
- T is homogeneous(T preserves scalar multiplication):
  for any v in V and any scalar k, $T(k * v) = k * T(v)$.

GRAPHICAL REPRESENTATION

EXAMPLE 1.5 Let T be a transformation defined by the matrix

```
> A:=matrix([[1,0],[0,-1]]);
```

Apply this transformation to the set of points given by the line $y = x + 1$

```
> S:={}: for i to 60 do t:=(i-1)/60:
```

```
  S:=S union {[t,t+1]}:od:
> BaseGeometry(A,S);
```

EXAMPLE 1.6   Let T be a transformation defined by the matrix

```
> A:=matrix([[0,1],[1,0]]);
```

Apply this transformation to the set of points on the graph of $y = x^2$

```
> S:={}: for i to 20 do t:=(i-1)/20:
  S:=S union {[t,t^2]}:od:
> BaseGeometry(A,S);
```

LEARNING THE PROCESS

Use the demonstration version of the function `lineartran` to learn the steps needed to verify that a transformation is linear. Here is an example.

EXAMPLE 1.7   Consider a transformation from $\mathbb{R}^2$ to $\mathbb{R}^3$

```
> T:=x->(x[1],x[2],0);
> lineartran(T,R2,R3);
```

Choose your own transformations and repeat this procedure until you have mastered it.

SOME PROPERTIES OF LINEAR TRANSFORMATIONS

Does the set of all linear transformation from a linear space V to a linear space W form a linear space?

EXAMPLE 1.8   Define two linear transformations from $\mathbb{R}^3$ to $\mathbb{R}^2$

```
> T1:=x->(x[1]-x[2],x[2]-x[3]); T2:=x->(x[2],x[1]+x[3]);
```

Define the sum T of the two transformations by $T(x) = (T_1 + T_2)(x) = T_1(x) + T_2(x)$

```
> T:=x->(x[1],x[1]+x[2]);
```

Is T a linear transformation? Check using the nostep mode of the `lineartran` function

```
> lineartran(T,R3,R2);
```

The sum of two linear transformations is linear.

Is the scalar multiple of a linear transformation a linear transformation? Define the transformation $T = k * T_1$ by $(k * T_1)(x) = k * T_1(x)$:

```
> T:=x->(k*x[1]-k*x[2],k*x[2]-k*x[3]);
```

Is $k * T_1$ a linear transformation? (If you want to use `lineartran`, choose different values of k; for example, $k = 1, k = 2, k = -2 \dots$.)

The set of all linear transformations from $\mathbb{R}^n$ to $\mathbb{R}^m$ forms a linear space. (Fact 5.1)

What is the image of the zero vector by a linear transformation?

EXAMPLE 1.9   Consider the transformation

```
> T:=x->(x[1]+x[2], x[2], 1);
```

Is T a linear transformation?

```
> lineartran(T,R2,R3);
```

What is the image of the zero vector of $\mathbb{R}^2$ under this transformation?

```
> 'T[0,0]' := [T(vector([0,0]))];
```

This example shows the image of the zero vector is not the zero vector under this transformation. Is T linear?

EXAMPLE 1.10   Consider the linear transformation

```
> T:=(x,y,z)->(x+y, y, y-z,z);
```

The image of the zero vector $[0,0,0]$ is

```
> 'T[0,0,0]' := [T(0,0,0)];
```

If T is a linear transformation, then the image of the zero vector is zero. (Fact 5.2)

Is the converse of this statement true? That is, if the image of the zero vector is the zero vector, is the transformation linear (Fact 5.2)?

If T is a linear transformation from vector space V into vector space W, and $B = \{v_1, v_2, \dots, v_n\}$ is a basis for V, is the image of the set $B = \{T(v_1), T(v_2), \dots, T(v_n)\}$ always a basis for a subspace of W?

EXAMPLE 1.11   Consider the linear transformation

```
> T:=(x,y,z)->(x+y,y,z);
```

Choose a basis for $\mathbb{R}^3$ consisting of $\{v_1, v_2, v_3\}$

```
> v1:=vector([1,-1,3]); v2:=vector([0,1,1]);
  v3:=vector([-3,1,0]);
```

The image of each basis vector is

```
> w1:=vector([T(1,-1,3)]); w2:=vector([T(0,1,1)]);
  w3:=vector([T(-3,1,0)]);
```

Does the set of vectors $\{w_1, w_2, w_3\}$ form a basis for $\mathbb{R}^3$ or a basis for a subspace of $\mathbb{R}^3$? Check whether the set is linearly independent by constructing the matrix whose rows are the w's and applying the `rref` to obtain the reduced echelon form

```
> A:=matrix([[0,-1,3],[1,1,1],[2,1,0]]);
> rref(A);
```

Is set $\{w_1, w_2, w_3\}$ linearly independent? The reduced echelon form shows that the vectors are linearly independent. Therefore, set $\{w_1, w_2, w_3\}$ forms a basis for $\mathbb{R}^3$. Hence, in this example, the image set $\{w_1, w_2, w_3\}$ of the basis set $\{v_1, v_2, v_3\}$ for $\mathbb{R}^3$ is a basis for $\mathbb{R}^3$.

EXAMPLE 1.12   Consider a linear transformation on $\mathbb{R}^4$

```
> T:=(x1,x2,x3,x4)->(x2+x1,x2,x3+x4,x4);
```

Choose a basis for $\mathbb{R}^4$

```
> v1:=vector([1,-1,3,0]); v2:=vector([-1,0,1,1]);
v3:=vector([-3,1,0,0]); v4:=vector([0,0,0,1]);
```

The image of each basis vector is

```
> w1:=vector([T(1,-1,3,0)]); w2:=vector([T(-1,0,1,1)]);
w3:=vector([T(-3,1,0,0)]); w4:=vector([T(0,0,0,1)]);
```

Is the set of vectors $\{w_1, w_2, w_3, w_4\}$ a basis for $\mathbb{R}^4$? Construct the matrix whose rows are the w's and apply `gausselim` or `rref`

```
> A:=matrix([[0,-1,3,0],[-1,0,2,1],
[-2,1,0,0], [0,0,1,1]]);
> rref(A);
```

What does the `rref(A)` tell us about the set $\{w_1, w_2, w_3, w_4\}$? This set is linearly independent and it is a basis for $\mathbb{R}^4$.

If T : V $\rightarrow$ W is a linear transformation between the linear spaces V and W and if B is a basis for V then the image of the elements of B is a basis for a subspace of W. (Fact 5.4)

OTHER EXAMPLES OF LINEAR TRANSFORMATIONS

Following are two examples of linear transformations used in calculus.

The first transformation is defined on the space of all continuous functions over a closed interval [a, b] (denote this space by C[a, b]). Elements of this space are continuous functions. From calculus, we know that the set C[a, b] is closed under addition $(f + g)(x) = f(x) + g(x)$ — and scalar multiplication $(kf)(x) = kf(x)$

EXAMPLE 1.13   Define the transformation

```
> T:=f->Int(f,x=a..b);
```

Does the transformation preserve the addition and scalar multiplication in C[a, b]? First check addition:

```
>  'T(f+g)':=T(f+g);'T(f)+T(g)':=T(f)+T(g);
```

Then check scalar multiplication

```
>  'T(k*f)':=T(k*f);  'k*T(f)':=k*T(f);
```

On comparing addition and scalar multiplication, what do we conclude? The transformation given by an integral defines a linear transformation.

The second transformation is defined on the space of all differential functions denoted by C′[a, b]. Elements of this space are differential functions. Again, from calculus we know that this set is closed under addition and scalar multiplication.

EXAMPLE 1.14   Define the transformation

```
>  T:=f->D(f);
```

Does this transformation preserve the addition and scalar multiplication in C′[a, b]? First check addition

```
>  'T(f+g)':=T(f+g);'T(f)+T(g)':=T(f)+T(g);
```

Then check scalar multiplication

```
>  'T(k*f)':=T(k*f);  'k*T(f)':=k*T(f);
```

On comparing addition and scalar multiplication, what do we conclude? The transformation defined by differentiation is a linear transformation.

EXERCISES

In the following exercises you may need to use the automated function `lineartran` and the Maple commands `matrix`, `vector`, `solve`, `evalm`, `gausselim`, and `rref`.

1. Consider the mapping

```
>  T:=(x,y,z)->(x,1,1);
```

Show that T is not an example of a linear transformation.

2. Consider the mapping

```
>  T:=x->(x[1]-x[2],x[2],0);
```

Verify that T is a linear transformation. You may use the interactive mode of `lineartran`.

3. Let $T : \mathbb{R}^2 \to \mathbb{R}^2$ be a linear transformation defined by the matrix

```
> A:=theta->matrix([[cos(theta),
> -sin(theta)],[sin(theta),cos(theta)]]);
```

Describe geometrically the action of the transformation T on the vector

```
> v:=vector([1,2]);
```

for the following values

```
> theta:= 0, Pi/3, Pi/4, Pi/2, 2*Pi/3, Pi;
```

4. Let $T : \mathbb{R}^3 \to \mathbb{R}^3$ be a linear transformation given by

```
> T:=(x,y,z)->(x-y,y+z,2*x-y+z);
```

and $S = \{v_1, v_2, v_3\}$ be a basis for the vector space $\mathbb{R}^3$

```
> v1:=vector([1,1,0]); v2:=vector([-1,0,1]);
> v3:=vector([0,0,1]);
```

   a. Show that for each u in $\mathbb{R}^3$, T(u) is a linear combination of the vectors $T(v_i)$, $i = 1, 2, 3$. (To compute $T(v_1)$ use: $T(v_1) = [T(1, 1, 0)]$;
   b. Does the set $\{T(u_1), T(u_2), T(u_3)\}$ span $\mathbb{R}^3$? Explain.

5. Let $T_1 : U \to V$ and $T_2 : V \to U$ be two linear transformations. Show that the composition $T_1 o T_2$ defined by $(T_1 o T_2)(v) = T_1(T_2(v))$ is a linear transformation (**Fact 5.3**).

Kernel and Range of a Linear Transformation

Two important subspaces associated with a linear transformation are the **kernel** and the **range**. These subspaces are fundamental in characterizing a linear transformation and in characterizing the solution set of any operator equation of the form Ax = b.

Recall that a nonhomogeneous system of linear equations Ax = b is consistent if and only if rank(A | b) = rank(A). Then the general solution is written

$$x = x_h + x_p$$

where x_h is a solution of the associated homogeneous system Ax = 0 and where x_p is the particular solution of the nonhomogeneous system Ax = b. The solution x_h belongs to the set {x | Ax = 0}. The particular solution x_p exists provided b belongs to the column space of A. The former set defines the **kernel** of a linear transformation; the latter set characterizes the **range** of a transformation.

Initialize the packages

```
> with(linalg):with(lintran);
```

DEFINING KERNEL AND RANGE

EXAMPLE 2.1   Given the matrices

```
> A:=matrix([[1,3,-1],[0,1,1],[1,4,0]]);
> x:=matrix([[x1],[x2],[x3]]);
```

define the linear transformation T by matrix multiplication

```
> T:=x->multiply(A,x);
```

Find all possible vectors x for which T(x) = 0

```
> T(x)=evalm(matrix([[0],[0],[0]]));
```

Construct the augmented matrix

```
> AUG:=matrix([[1,3,-1,0],[0,1,1,0],[1,4,0,0]]);
```

and apply Gauss elimination

> AUG1:=gausselim(AUG);

Is the system consistent? The system is consistent and the solution is given by

> solution:=backsub(AUG1);

Therefore the set of vectors x for which $T(x) = 0$ is the solution to the system $Ax = 0$. Every vector in this set is a multiple of $[4, -1, 1]$.

Now find all possible vectors y

> y :=matrix([[a],[b],[c]]);

for which there is a vector x such that $T(x) = y$.

> T(x)=evalm(y);

Construct the augmented matrix

> AUG:=matrix([[1,3,-1,a],[0,1,1,b],[1,4,0,c]]);

and apply Gauss elimination

> AUG1:=gausselim(AUG);

For the system to be consistent, the components of the vector $y = [a, b, c]$ must satisfy the relation $a + b = c$. Thus, for any vector y in the set $\{[a, b, c], a + b = c\}$, there is an x for which $T(x) = y$. In this case, the augmented matrix reduces to

> AUG1:=matrix([[1,3,-1,a],[0,1,1,b],[0,0,0,0]]);

and the solution of the system is

> solution:=backsub(AUG1);

The general solution can then be written

$$x = [4t_3 - 3b + a, -t_3 + b, t_3] = [4t_3, -t_3, t_3] + [-3b + a, b, 0]$$

$$= x_h + x_p$$

with $x_h = [4t_3, -t_3, t_3]$ being the solution of the associated homogeneous system $Ax = 0$ and $x_p = [-3b + a, b, 0]$ is the particular solution to $Ax = y$.

The first part of the example characterizes the **kernel** of a linear transformation and the second part is related to determining the **range** of a linear transformation.

EXAMPLE 2.2 Let T be the linear transformation from $\mathbb{R}^3$ to $\mathbb{R}^3$ defined by matrix multiplication

> A:=matrix([[1,3,-1],[1,0,1],[0,1,2]]);
x:=matrix([[x1],[x2],[x3]]);
> T:=x->multiply(A,x);

Find all possible vectors b

```
> b:=matrix([[b1],[b2],[b3]]);
```

for which there is a vector x such that $T(x) = b$

```
> T(x)=evalm(b);
```

As in Example 2.1, construct the augmented matrix and apply Gauss elimination

```
> AUG:=matrix([[1,3,-1,b1],[1,0,1,b2],[0,1,2,b3]]);
> AUG1:=gausselim(AUG);
```

This system is consistent and has a unique solution for any vector b in $\mathbb{R}^3$. The solution of the system in this case is given by $x = x_h + x_p$, where $x_h = [0, 0, 0]$ is the solution of the associated homogeneous system $Ax = 0$. The particular solution to $Ax = b$ x_p, is

```
> xp:=backsub(AUG1);
```

SUMMARY

Let $T : V \to W$ be a linear transformation between the two vector spaces V and W. The kernel of T, denoted by ker(T), is the set of all vectors v in V such that $T(v) = 0$; that is, ker(T) = {$v \in V \mid T(v) = 0$} The dimension of ker(T) or the nullity(T) is the number of elements in a basis for ker(T). **The kernel of T, ker(T), is a subspace of V. (Fact 5.5a)**

The range of T, denoted by range(T), is the set of all vectors w in W such that there is a vector v in V for which $T(v) = w$; that is, range(T) = {$w \in W \mid T(v) = w$ for some $v \in V$}. The dimension of range(T) or the rank(T) is the number of elements in a basis for range (T). **The range of T, range(T), is a subspace of W. (Fact 5.5b)**

LEARNING THE PROCESS

Use the demonstration version of `kernel` and `range` to learn how to construct these two sets. Here is an example.

EXAMPLE 2.3 Let T be the linear transformation from $\mathbb{R}^3$ to $\mathbb{R}^2$

```
> T:=x->(x[1],x[2]-x[3]);
> kernel(T,R3,R2);
```

Can you guess what the range of the transformation is?

```
> range(T,R3,R2);
```

Repeat using transformations of your own until you have learned how to construct the kernel and the range of a linear transformation.

SOME PROPERTIES OF KERNEL AND RANGE

EXAMPLE 2.4 Consider the linear transformation from $\mathbb{R}^3$ to $\mathbb{R}^2$

```
> T:=x->(x[1],x[3]-x[2]);
> kernel(T,R3,R2);
```

Can you guess what the range of the transformation is?

```
> range(T,R3,R2);
```

In this example the transformation $T : \mathbb{R}^3 \to \mathbb{R}^2$ has nullity(T) = 1 and rank(T) = 2. Compare the sum nullity(T) + rank(T) with the dimension of the domain $\mathbb{R}^3$.

EXAMPLE 2.5   Consider the linear transformation $T : \mathbb{R}^4 \to \mathbb{R}^3$

```
> T:=x->(x[1]-x[2],x[3]-x[4],0);
```

Use the nostep mode of `kernel`

```
> kernel(T,R4,R3);
```

Can you guess what the range of the transformation is? Use the nostep mode of `range`.

```
> range(T,R4,R3);
```

In this example, the transformation $T : \mathbb{R}^4 \to \mathbb{R}^3$ has nullity(T) = 2 and rank(T) = 2. Again, compare the sum nullity(T) + rank(T) to the dimension of the domain $\mathbb{R}^4$.

Dimension theorem Suppose V is a linear space of dimension n. If T is a linear transformation from V into W, then nullity(T) + rank(T) = n. (Fact 5.6)

EXAMPLE 2.6   Consider $T : \mathbb{R}^4 \to \mathbb{R}^4$

```
> T:=x->(x[1]-x[3],x[2]-x[4],x[3],x[4]);
```

Use the nostep mode of `kernel`

```
> kernel(T,R4,R4);
```

In this example, ker(T) = {0} and thus its nullity(T) = 0. From the information about the kernel, can you determine the range of the transformation? The dimension theorem gives us a way of figuring out the range from the kernel. The rank(T) = 4 and therefore the range of T is $\mathbb{R}^4$.

EXAMPLE 2.7   Consider the linear transformation $T : \mathbb{R}^4 \to \mathbb{R}^5$

```
> T:=x->(x[1]-x[4],x[2]-x[3],x[3]-x[5],x[4],x[5]);
```

Use the nostep mode of `kernel`

```
> kernel(T,R5,R5);
```

As in the previous example, ker(T) = {0}. Thus rank(T) = 5. Describe the range of T. The range of T is $\mathbb{R}^5$. What are those transformations whose kernel is the trivial subspace {0}?

A transformation T is one-to-one if and only if its kernel is the zero vector.

EXERCISES

In the following exercises you may need to use the automated functions `lineartran`, `kernel`, `range` and the Maple commands `matrix`, `vector`, `solve`, `evalm`, `gausselim`, and `rref`.

1. Let $T : \mathbb{R}^2 \to \mathbb{R}^2$ be a linear transformation given by matrix multiplication with the matrix

```
> A:=matrix([[2,-1],[-8,4]]);
```

   a. Describe the action of the linear transformation on any vector in $\mathbb{R}^2$.
   b. Does the vector

```
> v:=vector([1,2]);
```

   belong to ker(T)?
   c. Does the vector

```
> u:=vector([1,-4]);
```

   belong to range(T)?

2. Consider the linear transformation

```
> T:=(x,y,z)->(x,y,0);
```

   a. What does this transformation represent geometrically?
   b. Describe the kernel of this transformation.
   c. Describe the range of this transformation.

3. Consider the system of linear equations

```
> eq1:=x1 -x2 +2*x3 -x4 =11,;
> eq2:= 2*x1 +3*x2 -x3 +x4 =15;
> eq3:= 5*x1 +5*x3 -2*x4= a;
```

   a. Write the augmented matrix AUG and the coefficient matrix A of the system.
   b. Describe the row space of matrix AUG. Does it depend on the values of a?
   c. Describe the row space of matrix A. Does it depend on the values of a?
   d. Describe the column space of matrix AUG? Does it depend on the values of a?
   e. Describe the linear transformation T given by matrix multiplication with matrix A.
   f. Describe the ker(T). How does it relate to the row space of A?
   g. Describe the range(T). Does the vector

```
> v:=vector([11,15,a]);
```

   belong to range(T) for any A? How does this relate to the column space of AUG?

4. Describe all transformations $T : \mathbb{R}^n \to \mathbb{R}^m$ whose kernel is the space $\mathbb{R}^n$?

5. Let $T : U \to U$ be a linear transformation. Then the following are equivalent (Facts 5.7 and 5.8):

a. ker(T) = {0}.
b. T is one-to-one.
c. The inverse of T exists.

Matrix Representation of a Linear Transformation

An $m \times n$ matrix is an array of numbers with m rows and n columns. We shall show that a matrix is nothing but a linear transformation from one linear space into another. Is there a one-to-one correspondence between the set of all $m \times n$ matrices and the set of all linear transformations from $\mathbb{R}^n \to \mathbb{R}^m$?

Initialize the packages

```
> with(linalg): with(lintran);
```

MATRIX REPRESENTATION OF A LINEAR TRANSFORMATION

EXAMPLE 3.1   Let $T : \mathbb{R}^2 \to \mathbb{R}^3$ be a linear transformation given by

```
> T:=(x,y) -> (2*x,x+y,3*y);
```

Find the matrix representation of T with respect to the standard bases for $\mathbb{R}^2$ and $\mathbb{R}^3$.

1. Find the action of T on the standard basis of $\mathbb{R}^2$

```
> 'T(1,0)':=vector([T(1,0)]); 'T(0,1)':=vector([T(0,1)]);
```

2. Write each of the resulting vectors as a linear combination of the standard basis of $\mathbb{R}^3$

```
> 'T(1,0)':=2*evalm([1,0,0])+1*evalm([0,1,0])
+'0'*evalm([0,0,1]);
> 'T(0,1)':='0'*evalm([1,0,0])+1*evalm([0,1,0])
+3*evalm([0,0,1]);
```

3. The coordinates of vector $T(1,0)$ with respect to the standard basis are 2, 1, and 0. The coordinates of vector $T(0,1)$ with respect to the standard basis are 0, 1,

and 3. Construct the matrix whose columns consists of the coordinates of the vectors T(1, 0) and T(0, 1).

```
> A:=matrix([[2,0],[1,1],[0,3]]);
```

Thus the 3 × 2 matrix A is the matrix that represents the transformation T from $\mathbb{R}^2$ to $\mathbb{R}^3$. To check that this matrix represents the transformation T with respect to the standard basis, we multiply matrix A with the vector

```
> v:=matrix([[x],[y]]);
> A1:=multiply(A,v);
```

which coincides with the definition of the transformation T.

EXAMPLE 3.2 Let $T : \mathbb{R}^4 \to \mathbb{R}^3$ be a linear transformation given by

```
> T:=(x1,x2,x3,x4) -> (x1,x1+x2,x3+x4);
```

What is the matrix representation of T with respect to the standard bases for $\mathbb{R}^4$ and $\mathbb{R}^3$?

1. Find the action of T on the standard basis of $\mathbb{R}^4$:

```
> w1:=vector([T(1,0,0,0)]); w2:=vector([T(0,1,0,0)]);
> w3:=vector([T(0,0,1,0)]); w4:=vector([T(0,0,0,1)]);
```

2. Write the vectors $w_1, w_2, w_3,$ and w_4 as a linear combination of the standard basis of $\mathbb{R}^3$

```
> w1:=1*evalm([1,0,0])+1*evalm([0,1,0])+'0'
*evalm([0,0,1]);
> w2:='0'*evalm([1,0,0])+1*evalm([0,1,0])+'0'*
evalm([0,0,1]);
> w3:='0'*evalm([1,0,0])+'0'*evalm([0,1,0])+1*
evalm([0,0,1]);
> w4:='0'*evalm([1,0,0])+'0'*evalm([0,1,0])+1*
evalm([0,0,1]);
```

3. Construct the matrix whose columns consists of the coordinates of the w_i's with respect to the standard basis of $\mathbb{R}^3$

```
> A:=matrix([[1,0,0,0],[1,1,0,0],[0,0,1,1]]);
```

Therefore, the matrix that represents the transformation T from $\mathbb{R}^4$ to $\mathbb{R}^3$ is the 3 × 4 matrix A. Verify that this 3 × 4 matrix A represents the transformation $T : \mathbb{R}^4 \to \mathbb{R}^3$ with respect to the standard bases.

A linear transformation $T : \mathbb{R}^n \to \mathbb{R}^m$ can be represented as an $m \times n$ matrix whose columns are the coordinates of the vectors $w_i = T(e_i)$ with respect to the standard bases of $\mathbb{R}^m$ and $\mathbb{R}^n$. (Fact 5.9)

In general, if V and W are two finite dimensional linear spaces of dimension n and m respectively, then there is a one-to-one correspondence between the set of all linear transformations from V to W and the set of all $m \times n$ matrices.

LEARNING THE PROCESS

Use the demonstration mode of matrixrep to learn the process of finding the matrix representation of a linear transformation. Here are some examples

EXAMPLE 3.3   Let $T : \mathbb{R}^3 \to \mathbb{R}^3$ be the linear transformation given by

```
> T:=x->(x[1]- x[2],x[2]+x[3],x[3]);
> matrixrep(T,R3,R3);
```

EXAMPLE 3.4   Find the matrix representation of the linear transformation $T : \mathbb{R}^4 \to \mathbb{R}^3$

```
> T:=x->(x[1]- x[2]-x[3],x[2]+x[4],x[4]-2*x[3]);
> matrixrep(T,R4,R3);
```

Repeat using your own transformations until you have mastered the process of finding the matrix representation.

INVERTIBLE TRANSFORMATIONS

EXAMPLE 3.5   Let $T : \mathbb{R}^3 \to \mathbb{R}^3$ be a linear transformation given by

```
> T:=x->(x[1]- x[2],x[2]-x[3],x[3]- 2*x[2]);
```

Use the nostep mode to get the matrix representation of T

```
> matrixrep(T,R3,R3);
```

Thus the 3×3 matrix that represents T is

```
> A:=matrix([[1,-1,0],[0,1,-1],[0,-2,1]]);
```

Is A a nonsingular matrix?

```
> invA:=inverse(A);
```

What is the kernel of such a transformation? Use the nostep mode to compute the kernel

```
> kernel(T,R3,R3);
```

The transformation T whose kernel is {0} is an example of a one-to-one transformation. Such a transformation T has an inverse. How do we obtain the inverse of T? Choose any vector v in $\mathbb{R}^3$ and find the product of the inverse of A and v

```
> v:=matrix([[x1],[x2],[x3]]); invT:=multiply(invA,v);
```

Thus, the inverse of the transformation T is

```
> T1:=x->(x[1]-x[2]-x[3],-x[2]-x[3],-2*x[2]-x[3]);
```

SUMMARY

If T is a linear transformation $T : \mathbb{R}^n \to \mathbb{R}^n$, then $\ker(T) = \{0\}$

if and only if

T is a one-to-one transformation

if and only if

T has an inverse

if and only if

the matrix representing T is nonsingular

if and only if

the associated homogeneous system of linear equations $Ax = 0$

has only the trivial solution

if and only if

the associated nonhomogeneous system of linear equations

$Ax = b$ has a unique solution.

EXAMPLE FROM CALCULUS

EXAMPLE 3.6 Let P_1 be the set of all polynomials of degree less than or equal to 1 with basis $\{1, x\}$, and let P_2 be the set of all polynomials of degree less than or equal to 2 with basis $\{1, x, x^2\}$. Define $T : P_2 \rightarrow P_1$ by

```
> T:=x->(a-2*b)*x+4*c;
```

What is the matrix representation of T with respect to the given bases? As in the case of vectors, compute the image of each basis element of P_2. For example, compute the image of x^2 by setting $a = 1, b = 0$ and $c = 0$. Thus the basis of P_2 is transformed to $x^2 = ax^2 + bx + c$ with

```
> a:=1:b:=0:c:=0:'T(x^2)':=T(x);
```

$x = ax^2 + bx + c$ with

```
> a:=0:b:=1:c:=0:'T(x)':=T(x);
```

$1 = ax^2 + bx + c$ with

```
> a:=0:b:=0:c:=1:'T(1)':=T(x);
```

We now represent the images $x, -2x$ and 4 in terms of the basis of P_1 as

```
> x =1*x + '0'*'.1'; -2*x = -2*x +'0'*'.1';
4 = '0'*x + 4*'.1';
```

Thus, the 2×3 matrix that represents the transformation $T : P_2 \rightarrow P_1$ is

```
> A:=matrix([[1,-2,0],[0,0,4]]);
```

What is the image of the polynomial $2x^2 - 3x + 8$?

EXERCISES

In the following exercises you may need to use the automated functions `lineartran`, `kernel`, and `range`, `matrixrep` and the Maple commands `matrix`, `vector`, `solve`, `evalm`, `gausselim`, and `rref`.

1. Consider the linear transformation

```
> T:=(x1,x2,x3)->(x2-x1,0,x3-x1);
```

    a. Find the matrix representation of T with respect to the standard basis. Use the interactive mode of `matrixrep`.

    b. Find the rank and the nullity of T.

    c. Is the transformation T a one-to-one transformation? If so, find the inverse of T.

2. Let $T : \mathbb{R}^3 \to \mathbb{R}^3$ be the transformation

```
> T:=(x1,x2,x3)->(x1+x2, 2*x1 -2*x2, 2*x1-x2+x3);
```

    a. Find the matrix representation of transformation T.

    b. Find the nullity(T).

    c. Is this transformation invertible? If so, find its inverse.

3. Let T be a linear transformation defined on the set P_2 of all polynomials of degree less than or equal to 2 whose basis is $\{1, x, x^2\}$. Define $T : P_2 \to P_2$ by $T(ax^2 + bx + c) = \frac{d}{dx}(ax^2 + bx + c)$, where $\frac{d}{dx}$ is the derivative with respect to x.

    a. Find the matrix representation of T with respect to the given basis of P_2.

    b. What is the image of the polynomial $7x^2 + 13x - 8$?

4. Consider the linear transformations $T_1 : \mathbb{R}^3 \to \mathbb{R}^2$ and $T_2 : \mathbb{R}^2 \to \mathbb{R}^3$

```
> T1:=(x1,x2,x3)->(x1-x2,x2-x3);
  T2:=(x1,x2)->(x1+x2,x2,x2-x1);
```

    a. Find the matrix representation A_1 of T_1 and A_2 of T_2.

    b. Find the matrix representation of the composition of $T_1 o T_2$ and of $T_2 o T_1$ if defined.

Change of Basis

In Lesson 5.3 we described the matrix representation of a linear transformation with respect to the standard bases of the underlying linear spaces. How do we obtain the matrix representation of a transformation with respect to nonstandard bases? How do we change from a nonstandard basis to a standard one and conversely?

Initialize the packages

```
> with(linalg):with(lintran);
```

MATRIX REPRESENTATION WITH RESPECT TO NONSTANDARD BASIS

EXAMPLE 4.1 Consider the vector

```
> v:=vector([5,4]);
```

Choose two different bases for $\mathbb{R}^2$: the standard basis $E = \{e_1, e_2\}$

```
> e1:=vector([1,0]);e2:=vector([0,1]);
```

and a nonstandard basis $B = \{u_1, u_2\}$

```
> u1:=vector([3,2]);u2:=vector([1,1]);
```

With respect to standard basis E, the vector $v = [5, 4]$ can be easily represented

```
> [5,4]=5*evalm(e1)+4*evalm(e2);
```

Is it that easy to represent v with respect to basis B? That is, can you find, by inspection, c_1 and c_2 such that

```
> [5,4] = c1*evalm(u1) + c2*evalm(u2);
```

To determine c_1 and c_2, write the system of equations resulting from expressing v in terms of basis B. The related system of equations is

```
> eq1:=3*c1+c2 =5; eq2:=2*c1+c2=4;
```

Solve the system for c_1 and c_2

```
> solve({eq1,eq2},{c1,c2});
```

Thus, vector $v = [5, 4]$ with respect to basis B is

```
>  [5,4] =1*evalm(u1)+2*evalm(u2);
```

$c_1 = 1$ and $c_2 = 2$ are called the **coordinates** of vector v with respect to basis B.

The same procedure can be described in matrix form. Construct the matrix whose columns are the elements of basis B

```
>  S:=matrix([[3,1],[2,1]]);
```

Solve for the coordinates of vector v with respect to basis B using the matrix equation

```
>  c:=matrix([[c1],[c2]]): v:=matrix([[5],[4]]):
>  evalm(S)*evalm(c)=evalm(v);
```

Does the inverse of the matrix S exist? Why?

Since S is a nonsingular matrix, the coordinates of vector v with respect to basis B are obtained by multiplying the inverse of matrix S with vector v

```
>  c:=multiply(inverse(S),v);
```

EXAMPLE 4.2   Consider a vector in $\mathbb{R}^3$

```
>  v:=vector([1,-5,6]);
```

Choose two different bases for $\mathbb{R}^3$: the standard basis $E = \{e_1, e_2, e_3\}$

```
>  e1:=vector([1,0,0]); e2:=vector([0,1,0]);
   e3:=vector([0,0,1]);
```

and the basis $B = \{u_1, u_2, u_3\}$ where

```
>  u1:=vector([1,3,2]); u2:=vector([1,1,0]);
   u3:=vector([0,-1,3]);
```

With respect to standard basis E, vector $v = [1, -5, 6]$ can be easily represented:

```
>  [1,-5,6]=1*evalm(e1)+(-5)*evalm(e2)
   +6*evalm(e3);
```

Can you find c_1, c_2, and c_3 such that $v = c_1u_1 + c_2u_2 + c_3u_3$?

```
>  [1,-5,6] =c1*evalm(u1)+ c2*evalm(u2)+c3*evalm(u3);
```

To find c_1, c_2, and c_3, construct the related system of equations

```
>  eq1:=c1+ c2 =1; eq2:=3*c1+c2- c3=-5;
   eq3:=2*c1 +3*c3=6;
```

Solve the system for c_1, c_2, and c_3:

```
>  solve({eq1,eq2,eq3},{c1,c2,c3});
```

Vector $v = [1, -5, 6]$ with respect to basis B is expressed

```
>  [1,-5,6]=(-3/2)*evalm(u1)+
>  (5/2)*evalm(u2)+ 3*evalm(u3);
```

c_1, c_2 and c_3 are called the **coordinates** of the vector v with respect to the basis B.

We can obtain the same result by constructing the matrix whose columns are the elements of the basis

```
>  S:=matrix([[1,1,0],[3,1,-1],[2,0,3]]);
```

This computation is equivalent to writing the matrix equation $v = S * c$, where c is the coordinate vector of v with respect the basis B and v is the given vector. The solution is obtained by multiplying the inverse of matrix S with vector v:

```
>  c:=multiply(inverse(S),v);
```

Thus, v can be written:

```
>  evalm(v)=(-3/2)*evalm(u1)+(5/2)*evalm(u2)+3*evalm(u3);
```

To represent a vector with respect to a basis B other than the standard one involves the following steps.
    1. Construct the matrix S whose columns are the elements of the basis set B. This is called the transition matrix from basis B to the standard basis E.
    2. Construct the inverse of S.
    3. Multiply the inverse of S with the given vector to find its coordinates with respect to the basis B.

In general, to change the coordinates of a vector v with respect to a basis $B_1 = \{u_1, u_2, \ldots, u_n\}$ to a basis $B_2 = \{v_1, v_2, \ldots, v_n\}$:
    1. Construct the transition matrix S_1 from basis B_1 to the standard basis $B_1 \to S_1 \to E$.
    2. Construct the transition matrix S_2 from basis B_2 to standard basis $B_2 \to S_2 \to E$.
    3. Construct the transition matrix from basis B_1 to basis B_2 by multiplying the inverse of S_2 by S_1. (Facts 5.10 and 5.11)

EXAMPLE 4.3   Find the transition matrix corresponding to the change of basis from the basis $B_1 = \{v_1, v_2\}$

```
>  v1:=vector([5,2]); v2:=vector([7,3]);
```

to the basis $B_2 = \{u_1, u_2\}$

```
>  u1:=vector([3,2]); u2:=vector([1,1]);
```

The transition matrix from the basis B_1 to E is

```
>  S1:=matrix([[5,7],[2,3]]);
```

The transition matrix from the basis B_2 to E is

```
>  S2:=matrix([[3,1],[2,1]]);
```

Thus, the transition matrix from B_1 to B_2 is

```
> S:=multiply(inverse(S2),S1);
```

Now, if v is the vector $[74, 31]$ with respect to the standard basis E

```
> v:=vector([74,31]);
```

then its coordinates with respect to the basis B_1 are

```
> x:=multiply(inverse(S1),v);
```

Vector x with respect to basis B_2 is

```
> y:=multiply(S,x);
```

If S_1 and S_2 are the transition matrices from a basis B_1 to the standard basis E and from basis B_2 to E respectively, then $S = S_2^{-1}S_1$ is the transition matrix from B_1 to B_2. (Fact 5.12)

GEOMETRIC REPRESENTATION OF CHANGE OF BASIS

EXAMPLE 4.4   Consider an arc of a semicircle generated by

```
> A:={} : n:=100: for i to n do
t:=(i-1)/n: A:=A union {[t,sqrt(1-t^2)]}:od:
```

Consider the standard basis E and the basis $B = \{[1, 1], [1, -1]\}$. The transition matrix from B to E is

```
> S:=matrix([[1,1],[1,-1]]);
```

Display the image with respect to these bases.

```
> BaseGeometry(S,A);
```

EXAMPLE 4.5   Consider the curve generated by

```
> A:={} : n:=200: for i to n do
> t:=(i-1)/n: A:=A union {[t,t^2]}:od:
```

Consider the standard basis E and the basis $B = \{[1, 0], [0, -1]\}$. The transition matrix from B to E is:

```
> S:=matrix([[1,0],[0,-1]]);
```

Display the image with respect to these bases.

```
> BaseGeometry(S,A);
```

LEARNING THE PROCESS

Use the demonstration mode of `changebasis` to learn the process of representing a transformation with respect to different bases. Here is an example.

EXAMPLE 4.6   Consider the bases $G = \{v_1, v_2\}$ and $H = \{u_1, u_2\}$ of $\mathbb{R}^2$ given by

```
> v1:=vector([1,1]); v2:=vector([1,-1]);
> u1:=vector([-1,1]); u2:=vector([1,1]);
```

Construct the matrices whose columns are the vectors of the given bases

```
> G:=matrix([[1,1],[1,-1]]); H:=matrix([[-1,1],[1,1]]);
```

Find the transition matrices from basis G to basis H and from basis H to basis G.

```
> changebasis(G,H);
```

Repeat with your own bases until you have mastered the process of change of basis.

EXERCISES

In the following exercises you may need use the automated functions `lineartran`, `kernel`, `range`, and `changebasis` and the Maple commands `matrix`, `vector`, `solve`, `evalm`, `gausselim`, and `rref`.

1. Let $T : \mathbb{R}^4 \to \mathbb{R}^3$ be a linear transformation:

```
> T:=(x1,x2,x3,x4)->(x1+x2+x3,x2+x3,x3+x4);
```

a. Write the matrix representation for T with respect to the standard bases of $\mathbb{R}^4$ and $\mathbb{R}^3$.

b. Let $B = \{u_1, u_2, u_3, u_4\}$ be a basis for $\mathbb{R}^4$ given by

```
> u1:=vector([1,1,0,0]); u2:=vector([0,1,-1,0]);
> u3:=vector([0,1,0,-1]); u4:=vector([-1,0,0,1]);
```

and $B_1 = \{v_1, v_2, v_3\}$ be a basis for $\mathbb{R}^3$ given by

```
> v1:=vector([1,1,0]); v2:=vector([2,0,3]);
  v3:=vector([0,0,1]);
```

Find the matrix representation of T with respect to the basis B of $\mathbb{R}^4$ and B_1 of $\mathbb{R}^3$. Use this matrix to find the image of the vector $[3, 6, -9, 15]$.

2. Let $B = \{u_1, u_2, u_3, u_4\}$ be a basis for $\mathbb{R}^4$ given by

```
> u1:=vector([1,1,0,1]); u2:=vector([0,1,-1,0]);
> u3:=vector([0,1,0,-1]); u4:=vector([-1,0,0,1]);
```

a. Find the transition matrix from the standard basis of $\mathbb{R}^4$ to the basis B
Let $B_1 = \{v_1, v_2, v_3, v_4\}$ be another basis for $\mathbb{R}^4$ given by

```
> v1:=vector([1,1,1,1]);v2:=vector([1,1,-1,0]);
> v3:=vector([0,1,1,-1]);v4:=vector([-1,0,0,1]);
```

b. Find the transition matrix from the standard basis of $\mathbb{R}^4$ to the basis B_1.

c. Find the transition matrix from the basis B to the basis B_1.

Similarity

In this lesson we analyze the implication of the matrix representation of a linear transformation with respect to different bases.

Initialize the packages

```
> with(linalg):with(lintran);
```

SIMILARITY

EXAMPLE 5.1   Consider the linear transformation $T : \mathbb{R}^2 \to \mathbb{R}^2$

```
> T:=x->(x[1]+x[2],x[1]-x[2]);
```

Let $E = \{e_1, e_2\}$ be the standard basis for $\mathbb{R}^2$

```
> e1:=vector([1,0]); e2:=vector([0,1]);
```

Use the nostep mode of matrixrep to exhibit the matrix representation of T with respect to this basis

```
> matrixrep(T,R2,R2);
```

The matrix representation T_E with respect to the standard basis E is

```
> T(E):=matrix([[1,1],[1,-1]]);
```

Let $B = \{u_1, u_2\}$ be another basis for $\mathbb{R}^2$

```
> u1:=vector([1,2]); u2:=vector([2,1]);
```

Use the nostep mode of matrixrep to exhibit the matrix representation of T with respect to this basis

```
> matrixrep(T,R2,R2);
```

The matrix representation T_B with respect to the basis B is

```
> T(B):= matrix([[-5/3,-1/3],[7/3,5/3]]);
```

What is the relation between matrices T_E and T_B? The transition matrix from B to E is

```
> S:=matrix([[1,2],[2,1]]);
```

Compute the product $S^{-1}T_ES$

```
> 'inv(S).T(E).S':=multiply(inverse(S),multiply(T(E),S));
```

What is the relation between the matrices $S^{-1}T_ES$ and T_B? The matrix $S^{-1}T_ES$ is equal to T_B where T_E is the matrix representation with respect to the standard basis while T_B is the matrix representation with respect to the nonstandard basis B.

EXAMPLE 5.2   Consider the linear transformation $T: \mathbb{R}^3 \to \mathbb{R}^3$

```
> T:=x ->(x[1]+x[3],x[1]-x[2]+x[3],x[2]+x[3]);
```

Let $E = \{e_1, e_2, e_3\}$ be the standard basis for $\mathbb{R}^3$

```
> e1:=vector([1,0,0]); e2:=vector([0,1,0]);
e3:=vector([0,0,1]);
```

Use the nostep mode of `matrixrep` to exhibit the matrix representation of T with respect to this basis

```
> matrixrep(T,R3,R3);
```

The matrix representation T_E with respect to the standard basis is

```
> T(E):=matrix([[1,0,1],[1,-1,1],[0,1,1]]);
```

Let $B = \{u_1, u_2, u_3\}$ be another basis for $\mathbb{R}^3$

```
> u1:=vector([1,1,2]); u2:=vector([2,1,0]);
u3:=vector([0,1,1]);
```

Use the nostep mode of the `matrixrep` function to exhibit the matrix representation of T_B with respect to this basis

```
> matrixrep(T,R3,R3);
```

The matrix representation of T_B with respect to the basis B is

```
> T(B):= matrix([[5/3,2/3,5/3],[2/3,2/3,-1/3],
[-1/3,-1/3,-4/3]]);
```

What is the relation between the matrices T_E and T_B? The transition matrix from B to E is

```
> S:=matrix([[1,2,0],[1,1,1],[2,0,1]]);
```

Compute the product $S^{-1}T_ES$

```
> 'inv(S).T(E).S':=multiply(inverse(S),multiply(T(E),S));
```

What is the relation between the matrix $S^{-1}T_ES$ and T_B? The matrix $S^{-1}T_ES$ is equal to T_B where T_E is the matrix representation with respect to the standard basis while T_B is the matrix representation with respect to the nonstandard basis B.

Let $B_1 = \{v_1, v_2, v_3, \ldots v_n\}$ and $B_2 = \{u_1, u_2, u_3, \ldots u_n\}$ be two bases for a linear space V and T a linear transformation from V to V. If S is the transition matrix from the basis B_2 to B_1 and $T_{B_1)}$ is the matrix representing T with respect to the basis B_1, then the matrix $T_{B_2} = S^{-1}T_{B_1}S$ is the matrix representation of T with respect to the basis B_2. The matrix T_{B_2} is said to be **similar** to the matrix T_{B_1}.

In general, two matrices A and B are said to be similar if there is a nonsingular matrix S such that $B = S^{-1}AS$ (Fact 5.12)

LEARNING THE PROCESS

Use the function `matrixrep` to determine the matrix representation of a transformation with respect to different bases. Here is an example.

EXAMPLE 5.3 Consider the linear transformation T on $\mathbb{R}^3$

```
> T:=x->(x[1]+x[3],x[1]-x[2]+x[3],x[2]+x[3]);
```

Consider the two bases $G = \{v_1, v_2, v_3\}$ of $\mathbb{R}^3$ given by

```
> v1:=vector([1,1,0]); v2:=vector([0,1,1]);
  v3:=vector([0,0,1]);
```

and $H = \{u_1, u_2, u_3\}$ of $\mathbb{R}^3$ given by

```
> u1:=vector([1,1,4]); u2:=vector([3,1,1]);
  u3:=vector([-1,0,0]);
```

Find the matrix representation of T with respect to the bases G and H

```
> matrixrep(T,R3,R3);
```

Repeat with your own examples to master matrix representation.

EXERCISES

In the following exercises you may need to use the functions `lineartran`, `kernel`, `range`, `matrixrep`, and `changebasis` and the Maple commands `matrix`, `vector`, `solve`, `evalm`, `gausselim`, and `rref`.

1. Determine whether the following two matrices are similar

```
> A:=matrix([[1,0,0],[0,-1,0],[0,0,-1]]);
    B:=matrix([[-1,0,0],[0,-1,0],[0,0,1]]);
```

2. Let $T : \mathbb{R}^4 \to \mathbb{R}^3$ be a linear transformation

```
> T:=x->(x[1]+x[2]+x[3],x[2]+x[3],x[3]+x[4]);
```

a. Let $B = \{u_1, u_2, u_3, u_4\}$ be a basis for $\mathbb{R}^4$ given by

```
> u1:=vector([1,1,0,0]);u2:=vector([0,1,-1,0]);
  u3:=vector([0,1,0,-1]);u4:=vector([-1,0,0,1]);
```

Find the matrix A that represents T with respect to basis B

b. Let $B_1 = \{v_1, v_2, v_3, v_4\}$ be another basis for $\mathbb{R}^4$ given by

```
> v1:=vector([1,1,1,1]);v2:=vector([1,1,-1,0]);
  v3:=vector([0,1,1,-1]);v4:=vector([-1,0,0,1]);
```

Find the matrix A that represents T with respect to basis B_1.

c. Are matrices A and A_1 similar? Are the powers of A and A_1 similar?

Kernel and Range of a Linear Transformation

Purpose

The purpose of this lab is to study the kernel and range properties of a linear transformation and their role in the solution of systems of linear equations.

Automated Linalg functions

In this lab you will use the automated functions lineartran, matrixrep, kernel, range, and changebasis. To get help with a function, type at the Maple prompt >?function name; for example, >?kernel;

Instructions

1. To execute a statement move the cursor to the line using the mouse or the key board and press the enter key.
2. While executing a function, create Maple input regions if needed. Otherwise the output will not appear in the desired place.
3. Execute the following commands to load the package

   ```
   > with(linalg):with(lintran);
   ```

TASK 1

The purpose of this task is to verify whether a given transformation is linear or not.

Activity 1

Given the transformation T: $\mathbb{R}^3 \to \mathbb{R}^3$

```
> T:=x->(x[1]-x[2],x[3]-x[2],x[1]-x[3]);
```

T is a linear transformation on $\mathbb{R}^3$? Use the interactive mode of lineartran.

```
> lineartran(T, R3, R3);
```

Activity 2

Let T be a transformation from $\mathbb{R}^3$ to $\mathbb{R}^4$

```
> T:=x->(x[1],x[2],x[3],x[2]^2);
```

Is T is a linear transformation from $\mathbb{R}^3$ to $\mathbb{R}^4$? Use the interactive mode of lineartran.

Activity 3

What conditions must a transformation T satisfy to be a linear transformation?

TASK 2

The purpose of this task is to understand and determine the kernel and range of a linear transformation. Let T be a linear transformation from $\mathbb{R}^3$ to $\mathbb{R}^4$

```
> T:=x->(x[1]-2*x[2],x[2]-x[3],x[3]-x[2],x[2]-x[3]);
```

Activity 1

Enter the system of linear equations corresponding to $T(x) = 0$.

Activity 2

Solve the resulting homogeneous system of linear equations. What does the solution set represent?

Activity 3

Determine a basis for this solution set. What is its dimension? What is rank(T)? Can you describe the range(T)?

TASK 3

Let T be a linear transformation from $\mathbb{R}^3$ to $\mathbb{R}^4$. Assume that $\mathbb{R}^3$ and $\mathbb{R}^4$ are equipped with the standard bases. The action of T on the standard basis of $\mathbb{R}^3$ can be described as

```
> T(e1):=vector([1,1,1,-1]); T(e2):=vector([0,1,0,-2]);
> T(e3):=vector([5,7,1,3]);
```

Activity 1

Give the general description of T; that is, find T(x) for any vector $x = [x_1, x_2, x_3]$ in $\mathbb{R}^3$?

Activity 2

What is the kernel of T? Use, if you like, the interactive mode of the kernel function. What is the nullity of T?

Activity 3

Describe the range of T. What is rank(T)?

Activity 4

Using the information of this task, can you describe the nonhomogeneous systems of linear equations that are consistent?

EXTRA LAB PROBLEM

Let A be a 4×5 matrix given by:

```
> A:=matrix([[1,-1,0,2,3],[2,0,1,-2,4],[-3,4,1,0,5],
[0,3,2,0,12]]);
```

and

```
> x:=matrix([[x[1]],[x[2]],[x[3]],[x[4]],[x[5]]);
```

Let T the linear transformation from $\mathbb{R}^5$ to $\mathbb{R}^4$ be given by matrix multiplication

```
> T := x -> multiply(A,x);
```

1. Determine the kernel of T. How does the kernel of T relate to the row space of the matrix A? What does the information about the kernel imply about the solution of the associated homogeneous system $Ax = 0$?
2. Determine the range of T. How does the range of T relate to the column space of the matrix A? What does the information about the range imply about the solution of the associated nonhomogeneous system $Ax = b$?

Matrix Representation of Linear Transformations

Purpose

The purpose of this lab is to find the matrix representation of a linear transformation with respect to a given basis with respect to different bases and similar matrices.

Automated Linalg functions

In this lab you will use the automated functions `lineartran`, `matrixrep`, `kernel`, `range`, and `changebasis`. To get help with a function type at the Maple prompt >?function name; for example, >?kernel?

Instructions

1. To execute a statement move the cursor to the line using the mouse or the key board and press the enter key.
2. While executing a function, create Maple input regions if needed; otherwise the output will not appear in the desired place.
3. Execute the following commands to load the packages

   ```
   > with(linalg):with(lintran);
   ```

TASK 1

The purpose of this task is to obtain the matrix representation of a linear transformation with respect to the standard basis. Let T: $\mathbb{R}^2 \to \mathbb{R}^3$ be a linear transformation

```
> T:= x->(x[1]+x[2],x[2]-2*x[1],x[1]+3*x[2]);
```

Activity 1

What is the action of T on the basis $E_1 = \{[1,0],[0,1]\}$?

Activity 2

Represent the vectors of Activity 1 in terms of the basis $E_2 = \{[1,0,0],[0,1,0],[0,0,1]\}$ of $\mathbb{R}^3$.

Activity 3

Write the matrix that represents the transformation T with respect to the standard bases for $\mathbb{R}^2$ and $\mathbb{R}^3$.

Activity 4

In your own words, describe the process of obtaining the matrix representation of a linear transformation with respect to the standard basis.

Activity 5

Use the interactive mode of `matrixrep` to obtain the matrix representation of the transformation

```
> T:= x->(x[1]+x[2],x[2]-2*x[1],x[1]+3*x[2]);
```

TASK 2

The purpose of this task is to find the matrix representation of a linear transformation with respect to bases of the underlying vector spaces other than the standard ones. Let T be linear transformation from $\mathbb{R}^3$ to $\mathbb{R}^4$

```
> T:=x->(x[2],x[1],x[3]-x[2],x[1]+x[2]-x[3]);
```

Activity 1

What is the action of T on the basis $B_1 = \{[1,0,1],[0,1,1],[-1,1,2]\}$ for $\mathbb{R}^3$?

Activity 2

Represent the result of Activity 1 in terms of the basis

$$B_2 = \{[1,0,0,1],[1,1,1,0],[1,0,0,0],[1,-1,2,0]\}$$

of $\mathbb{R}^4$.

Activity 3

1. Use the result of Activity 2 to give the matrix representation of T.
2. Use the interactive mode of `matrixrep` to obtain the matrix representation of the transformation

```
> T:=x->(x[2],x[1],x[3]-x[2],x[1]+x[2]-x[3]);
```

TASK 3

The purpose of this task is to enforce the notion of the change of basis. Let $B = \{[1,1,1],[1,2,2],[2,3,4]\}$ be a basis for $\mathbb{R}^3$.

Activity 1

Find the transition matrix S from the standard basis of $\mathbb{R}^3$ to B.

Activity 2

What does the inverse of matrix S represent?

Activity 3

What are the coordinates of the vector $v = [3,6,9]$ with respect to basis B?

Activity 4

Let T be a linear transformation on $\mathbb{R}^3$

```
> T:=x->(x[1]+x[2],x[2]-x[1],x[1]+x[2]-x[3]);
```

1. What is the matrix representation of T with respect to the standard basis of $\mathbb{R}^3$?
2. What is the matrix representation of T with respect to the basis B?
3. Let $B_1 = \{[1,1,0],[-1,0,1],[0,0,1]\}$ be another basis for $\mathbb{R}^3$. How do we obtain the matrix representation with respect to B_1?
4. Are the two matrices representing T with respect to the bases B and B_1 similar?

EXTRA LAB PROBLEM

Let A and B be two matrices

```
> A:=matrix([[4,1,-1],[-2,5,2],[4,-4,1]]);
  B:=matrix([[3,3,1],[-1,-1,-1],[1,10,6]]);
```

1. Is there a 3×3 nonsingular matrix P such that $AP = PB$?

Consider the matrix

```
> C:=matrix([[3,1,0],[-1,3,1],[1,1,2]]);
```

2. Is there a 3×3 nonsingular matrix P such that $BP = PC$?
3. Compare the determinants of B and C. If they are equal, can you tell why?
4. In general, what is the relation between the determinants of similar matrices?
5. Are the powers of B^n and C^n similar? Experiment with $n = 2, 3$, and 4. Can you generalize this result? Prove your generalization.

Balancing Chemical Reactions

Purpose

The purpose of this application is to make use of linear transformations and their properties in a variety of settings.

Initialize the packages

```
> with(linalg):with(lintran);
```

CHEMICAL REACTION 1

Consider a chemical reaction in which hydrazine (N_2H_4) and dinitrogin tetraoxide (N_2O_4) combine to form nitrogen (N_2) and water(H_2O):

$$N_2H_4 + N_2O_4 \rightarrow N_2 + H_2O$$

One problem is the balancing of the right side and the left side of this chemical reaction: the number of atoms of each element on both sides must be the same.

Activity 1

Write the system of linear equations that balances the two sides. Explain what each variable stands for.

Activity 2

Describe the linear transformation that is represented by the system.

Activity 3

What is the kernel of this linear transformation?

Activity 4

Using the information about the kernel, can you balance the chemical reaction? Write the balanced chemical reaction.

CHEMICAL REACTION 2

Oxygen masks for producing oxygen O_2 in emergency situations contain potassium superoxide KO_2. It reacts with carbon dioxide and water H_2O in exhaled air to produce oxygen according to the reaction:

$$KO_2 + H_2O + CO_2 \rightarrow KHCO_3 + O_2$$

Activity 1

Write the system of linear equations that balances the two sides. Explain what each variable stands for.

Activity 2

Describe the linear transformation that is represented by the system.

Activity 3

What is the kernel of this linear transformation?

Activity 4

Using the information about the kernel, can you balance the chemical reaction? Write the balanced chemical reaction.

Management Science: Warehouse Problem

Purpose

This management science application deals with the problem of distribution of produce from warehouses into grocery stores.

Initialize the packages

```
> with(linalg):with(lintran);
```

DESCRIPTION OF WAREHOUSE PROBLEM

A produce company rents two warehouses. The company can store 16 and 12 tons of produce in warehouses W_1 and W_2, respectively. The company distributes the produce on a regular basis to two major grocery stores S_1 and S_2. Suppose that stores S_1 and S_2 can store A and b tons, respectively.

Activity 1

Write the system of linear equations $Ax = B$ that describes the quantities distributed from each warehouse to the stores. The constraints are that the total amounts stored in the warehouses must be distributed to the stores.

Activity 2

Describe the system of Activity 1 in terms of a transformation $T(x) = Ax$. Identify the elements x and A of this transformation. Interpret the meaning of $T(x) = B$.

Activity 3

Determine the kernel of transformation T.

Activity 4

What does Activity 3 imply about the values of matrix B? For example, does there exist a solution for every B in $\mathbb{R}^4$? If not, find the values of A and b so that the equation $T(x) = B$ is consistent. Explicitly state the conditions on A and b to achieve consistency.

Activity 5

Using the values of A and b in Activity 4, describe the feasible integer solutions to the problem $T(x) = B$.

Activity 6

The cost of shipping one ton of the produce to the respective store is

	S_1	S_2
W_1	80	45
W_2	60	55

Write the cost function C for distributing from the warehouses to the two stores.

Activity 7

Find the solution that renders the lowest cost.

Eigenspaces

Many problems in mathematics, engineering, physics, and other disciplines can be cast as eigenvalue problems of the form $Ax = \lambda x$. The idea is to find the nonzero solutions x associated with the scalar λ for this system of equations. We will use eigenvalue problems in characterizing the global extrema of functions and in studying quadratic forms.

Eigenvalues and Eigenvectors

The problem of finding nonzero solutions x to the matrix equation $Ax = \lambda x$ arises in the formulation of many problems in engineering and the physical sciences. The differential equation $y''(t) = -\lambda y$ that models the spring oscillation, with y being the displacement from the equilibrium position and λ the spring constant, is an example of such a problem, with A being a second-order linear differential transformation. This is called the **eigenvalue problem**. The nonzero vector x is called the **eigenvector** corresponding to the **eigenvalue** λ.

Initialize the following packages

```
> with(linalg):with(lineign);
```

WHAT ARE EIGENVALUES AND EIGENVECTORS?

EXAMPLE 1.1   Let A be a 3 × 3 matrix

```
> A:=matrix([[1,1,1],[0,2,1],[0,0,3]]);
```

x be a vector

```
> x:=matrix([[x1],[x2],[x3]]);
```

and Id be the 3 × 3 identity matrix

```
> Id:=diag(1,1,1);
```

The eigenvalue problem reduces to finding all λ's and the corresponding nonzero vectors x satisfying the equation

```
> evalm(A)*evalm(x) = lambda*evalm(x);
```

This is equivalent to solving the matrix equation

```
> (evalm(A)-lambda*evalm(Id))*evalm(x)=
matrix([[0],[0],[0]]);
```

or

```
> S:=evalm((A-lambda*Id)*(x))=matrix([[0],[0],[0]]);
```

The resulting system is

```
> eq1:=(1-lambda)*x1+x2+x3=0; eq2:=(2-lambda)*x2+x3=0;
eq3:=(-3+lambda)*x3=0;
```

When does this homogeneous system possess a nontrivial solution? Recall that a homogeneous system has nontrivial solutions if and only if the determinant of the coefficient matrix is equal to zero. That is, the coefficient matrix

```
> A1:=evalm(A-lambda*Id);
```

must have a zero determinant

```
> det(A1)=0;
```

The resulting equation is called the **characteristic polynomial** of A (the Maple command is `charpoly(A,lambda);`). The solution set of the characteristic polynomial is

```
> s:={solve(det(A1)=0,lambda)};
```

The elements of this solution set are the **eigenvalues** of A

```
> lambda1:=s[1]; lambda2:=s[2]; lambda3:=s[3];
```

How do we compute the eigenvectors, that is, the nonzero vectors associated with each eigenvalue? To compute the eigenvectors, we substitute the eigenvalues in the matrix equation S.

An eigenvector associated with the eigenvalue 1 is the solution of the system

```
> E1:=subs(lambda=s[1],S);
```

The system E_1 implies that $x_2 = x_3 = 0$ and x_1 is arbitrary. An eigenvector corresponding to the eigenvalue 1 is

```
> evector1:=vector([1,0,0]);
```

An eigenvector associated with the eigenvalue 2 is the solution of the system

```
> E2:=subs(lambda=s[2],S);
```

The system E_2 implies that $x_3 = 0$ and $x_1 = x_2$. Thus, an eigenvector corresponding to the eigenvalue 2 is

```
> evector2:=vector([1,1,0]);
```

An eigenvector associated with the eigenvalue 3 is the solution of the system

```
> E3:=subs(lambda=s[3],S);
```

The system E_3 implies that $x_1 = x_2 = x_3$. Thus, an eigenvector corresponding to the eigenvalue 3 is

```
> evector3:=vector([1,1,1]);
```

Matrix A has three distinct eigenvalues and three independent eigenvectors.

EXAMPLE 1.2   Let us consider another 4×4 matrix

```
> A:=matrix([[1,1,1,1],[0,2,0,0],[1,0,3,1],[0,0,0,1]]);
```

together with the identity matrix

```
> Id:=diag(1,1,1,1);
```

and the vector x

```
> x:=matrix([[x1],[x2],[x3],[x4]]);
```

We want to find all λ's and the corresponding nonzero vectors x satisfying the equation

```
> evalm(A)*evalm(x) = lambda*evalm(x);
```

This is equivalent to solving the matrix equation

```
> (evalm(A)-lambda*evalm(Id))*evalm(x)=
matrix([[0],[0],[0],[0]]);
```

or

```
> S:=evalm((A-lambda*Id)*(x))=matrix([[0],[0],[0],[0]]);
```

When does this homogeneous system possess a nontrivial solution? This matrix equation has a nonzero solution x provided the coefficient matrix

```
> A1:=evalm(A-evalm(lambda*Id));
```

has a zero determinant

```
> det(A1)=0;
```

We solve the characteristic polynomial of A to obtain the eigenvalues. The solution set is

```
> s:={solve(det(A1)=0,lambda)};
```

The eigenvalues are

```
> lambda1:=s[1]; lambda2:=s[2]; lambda3:=s[3];
lambda4:=s[4];
```

To get an eigenvector associated with the eigenvalue 1, we solve the system

```
> E1:=subs(lambda=1,S);
```

We get $x_1 = x_4, x_2 = 0, x_3 = -x_4$, and x_4 is arbitrary. Thus, an eigenvector corresponding to eigenvalue 1 is

```
> ev1:=vector([1,0,-1,1]);
```

To get an eigenvector associated with eigenvalue 2, we solve the system

```
> E2:=subs(lambda=2,S);
```

We get $x_1 = -x_3, x_2 = -2x_3, x_4 = 0$, and x_3 is arbitrary. Thus, an eigenvector associated with the eigenvalue 2 is

```
> ev2:=vector([-1,-2,1,0]);
```

To get an eigenvector associated with eigenvalue $2 + \sqrt{2}$, we solve the system

```
> E3:=subs(lambda=2+sqrt(2),S);
```

We get $x_1 = (-1 + \sqrt{2}) x_3$, $x_2 = 0$, $x_4 = 0$, and x_3 is arbitrary. Thus, an eigenvector associated with the eigenvalue $2 + \sqrt{2}$ is

```
> ev3:=vector([-1+sqrt(2),0,1,0]);
```

To get an eigenvector associated with eigenvalue $2 - \sqrt{2}$, we solve the system

```
> E4:=subs(lambda=2-sqrt(2),S);
```

We get $x_1 = (-1 - \sqrt{2}) x_3$, $x_2 = 0$, $x_4 = 0$, and x_3 is arbitrary. Thus, an eigenvector associated with the eigenvalue $2 - \sqrt{2}$ is

```
> ev4:=vector([-1-sqrt(2),0,1,0]);
```

Again, matrix A has four distinct eigenvalues and four independent eigenvector.

SUMMARY

λ is an eigenvalue of an $n \times n$ matrix A
if and only if
$(A - \lambda * \mathrm{Id}) * x = 0$ has a nontrivial solution
if and only if
$A - \lambda * \mathrm{Id}$ is a singular matrix
if and only if
$\det(A - \lambda * \mathrm{Id})$ is equal to zero.

With each eigenvalue λ, we associate an eigenspace E_λ. This space consists of all vectors spanned by the linearly independent eigenvectors associated with the eigenvalue. (Fact 6.1)

In Example 1.1, for the matrix

```
> A:=matrix([[1,1,1],[0,2,1],[0,0,3]]);
```

the eigenspace E_1 associated with eigenvalue 1 consists of all vectors of the form $[x_1, 0, 0]$ with x_1 an arbitrary scalar. The set consisting of the eigenvector $[1, 0, 0]$ is a basis for the eigenspace E_1.

For the eigenvalue 2, the eigenspace E_2 consists of all vectors of the form $[x_1, x_1, 0]$ with x_1 an arbitrary scalar. The set consisting of the eigenvector $[1, 1, 0]$ is a basis for this eigenspace.

For the eigenvalue 3, we associate the eigenspace E_3 consists of all vectors of the form $[x_1, x_1, x_1]$ with x_1 an arbitrary scalar. The set consisting of the eigenvector $[1, 1, 1]$ is a basis for this eigenspace.

LEARNING THE PROCESS

Use the demonstration mode of `eigenvals` and `eigenvects` to learn the process of obtaining the eigenvalues and eigenvectors of a given matrix. Start with the matrix:

> `A:=matrix([[-2,0,1],[-6,-2,0],[0,0,3]]);eigenvals(A);`

To compute the eigenvectors use the function `eigenvects(A)`.

> `eigenvect(A);`

Repeat using your own matrices until you have mastered the process.

SOME PROPERTIES OF EIGENVALUES AND EIGENVECTORS

EXAMPLE 1.3   Consider the matrix

> `A:=matrix([[1,1,1,1],[0,2,1,4],[0,0,3,8],[0,0,0,6]]);`

What is a relation between the sum of the eigenvalues and the sum of the diagonal elements of the matrix A?

Using the nostep mode of the function `eigenvals(A)`,

> `eigenvals(A);`

the eigenvalues of matrix A are 1, 2, 3, and 6. The sum of the eigenvalues is 12. Compare this to the sum of the diagonal elements of the matrix A

> `A[1,1]+A[2,2]+A[3,3]+A[4,4];`

EXAMPLE 1.4   Consider another matrix

> `A:=matrix([[1,2,3,4],[4,1,2,3],[3,4,1,2],[2,3,4,1]]);`

Using the nostep mode of the function

> `eigenvals(A);`

the eigenvalues of matrix A are $10, -2, -2 + 2 * I$ and $-2 - 2 * I$ $(I = \sqrt{-1})$. The sum of the eigenvalues is 4. Compare this to the sum of the diagonal elements of matrix A

> `A[1,1]+A[2,2]+A[3,3]+A[4,4];`

The sum of the eigenvalues of a matrix A is equal to the trace of A where the trace of a matrix is the sum of the diagonal elements.

EXAMPLE 1.5   Consider the matrix

```
> A:=matrix([[1,1,1],[1,1,1],[1,1,1]]);
```

What is the relation between the determinant of matrix A and the product of the eigenvalues?
Using the nostep mode to compute

```
> eigenvals(A);
```

The eigenvalues are 0, 0, and 3. Their product is 0. What is the determinant of the matrix? Since two rows of matrix A are identical, the determinant of A is 0.
Consider another matrix

```
> A:=matrix([[2,-1,2,3],[0,1,2,4],[3,1,9,1],[0,0,0,1]]);
```

Using the nostep mode to compute

```
> eigenvals(A);
```

the eigenvalues are $1, 2, 5 + 2\sqrt{6}, 5 + 2\sqrt{6}$. The product of the eigenvalues is 2. Now the determinant of the matrix A is

```
> det(A);
```

The product of the eigenvalues of a matrix A is equal to the determinant of the matrix.

EXERCISES

In the following exercises you may need to use the automated functions `eigenvals` and `eigenvects`, and the Maple commands `solve`, `matrix`, `vector`, `gausselim`, `rref`, and `charpoly`.

1. Consider the matrix:

```
> A:=matrix([[-15,16,32],[-4,5,8],[-4,4,9]]);
```

Use the interactive modes of the eigenvals and eigenvects functions to obtain the eigenvalues and eigenvectors of the matrix A.

2. Determine the eigenspaces associated with each eigenvalue of the matrix

```
> A:=matrix([[5,0,1],[1,1,0],[-7,1,0]]);
```

3. Consider the matrix

```
> A:=matrix([[1,0],[-1,2]]);
```

a. Define the characteristic polynomial of matrix A as

```
> p:=lambda->expand(charpoly(A,lambda));
```

Show that matrix A satisfies $p(A) = 0$.

b. Repeat part (a) using the matrix

```
> A:=matrix([[1,-1,1],[0,1,2],[0,1,-1]]);
```

c. Propose a general statement relating matrix A to its characteristic polynomial?

4. a. Given the circulant matrix of order 3

```
> A:=matrix([[1,2,3],[3,1,2],[2,3,1]]);
```

Show, using only matrix multiplication, that the vector

```
> v:=vector([1,1,1]);
```

is an eigenvector with eigenvalue 6. Verify your answer by using `eigenvects`.

b. Given the circulant matrix of order 4

```
> A:=matrix([·[1,2,3,4],[4,1,2,3],[3,4,1,2],[2,3,4,1]]);
```

Show, using only matrix multiplication, that the vector

```
> v:=vector([1,1,1,1]);
```

is an eigenvector with eigenvalue 10. Verify your answer by using `eigenvects`.

c. Given the circulant matrix of order 5

```
> A:=matrix([[1,2,3,4,5],[5,1,2,3,4],[4,5,1,2,3],
    [3,4,5,1,2],[2,3,4,5,1]]);
```

Show, using only matrix multiplication, that the vector $[1, 1, 1, 1, 1]$ is an eigenvector with eigenvalue 15.

d. If A is an $n \times n$ circulant matrix, is the vector $[1, 1, 1, \ldots, 1]$ an eigenvector? What is the associated eigenvalue? Verify your assertion.

5. The eigenvalues of a matrix A are $-1, 0$, and 1, and the corresponding eigenvectors are x_0, x_1, and x_2, respectively. Find the solution x to the system $Ax = x_0 + x_2$. Does the system $Ax = x_1$ have a solution? Verify your answer.

6. An investor wants to open three different accounts A_1, A_2, and A_3 with equal amounts. The accounts yield a yearly profit of 6%, 8%, and 10%, respectively. At the end of each year, the investor's policy is to invest $\frac{1}{3}$ of the money earned in A_2 and $\frac{2}{3}$ of the money earned in A_3 in A_1, and $\frac{1}{3}$ of the money earned in A_3 in A_2.

a. Write the system of equations that represents the amount invested in each account after n years.

b. Write the matrix representation of the system.

c. Express the amount of money in each account in year n in terms of the initial amount invested in each account *Hint*: Compute the eigenvalues and the eigenvectors of the coefficient matrix.

d. Using the result of part (c), estimate the number of years it takes to double the amount in A_1.

Matrix Diagonalization

If D is a given $n \times n$ diagonal matrix and b is an $n \times 1$ matrix, then it is easy to solve the matrix equation Dx = b for the unknown $n \times 1$ vector x. Suppose we are interested in solving the matrix equation Ax = b. If matrix A is **similar** to a diagonal matrix D (that is, there is a nonsingular matrix P such that $P^{-1}AP = D$), then Ax = b can be transformed to Dy = P^{-1}b that can be easily solved for y where y = P^{-1}x. Then x can be determined as x = Py. Thus, we are interested in the question: when is a given $n \times n$ matrix A similar to a diagonal matrix?

Initialize the packages

```
> with(linalg):with(lineign);
```

DIAGONALIZATION PROCESS

EXAMPLE 2.1 Let A be a 3×3 matrix

```
> A:=matrix([[0,1,-1],[1,0,1],[1,-1,-4]]);
```

The eigenvalues of matrix A are (use the nostep mode)

```
> eigenvals(A);
```

The corresponding eigenvectors are

```
> eigenvects(A);
```

The eigenvectors are

```
> v1:=vector([1,1,0]); v2:=vector([-1,1,-1]);
v3:=vector([-1,1,-2]);
```

Construct the matrix whose columns are eigenvectors v_1, v_2, and v_3 of matrix A

```
> P:=transpose(matrix([v1,v2,v3]));
```

Since the 3×3 matrix A has three independent eigenvectors, we can construct the matrix P whose columns are the eigenvectors. Is P a nonsingular matrix?

```
> det(P);
```

What if we perform the matrix multiplication: $P^{-1}AP$?

```
> P^('-1')*A*P = multiply(inverse(P),multiply(A,P));
```

In this case matrix A is similar to a diagonal matrix. This process is referred to as the **diagonalization process.**

LEARNING THE PROCESS

Use the demonstration mode of `diagonalize` to learn the process of diagonalizing a matrix. Remember that if the roots of the characteristic polynomial are not "nice," then the output may not be "appealing to the eye."

EXAMPLE 2.2   Is matrix A similar to a diagonal matrix?

```
> A:=matrix([[1,2,0,6],[-1,1,2,0],[3,4,1,2],[2,1,0,0]]);
> diagonalize(A);
```

Since the 4×4 matrix A has four independent eigenvectors, we can construct the matrix P whose columns are the eigenvectors. In this case, matrix A is similar to a diagonal matrix.

Repeat using your own matrices to learn the diagonalization process.

CONDITIONS FOR DIAGONALIZATION

Can we still diagonalize a matrix A if its eigenvalues are not distinct?

EXAMPLE 2.3   Consider the matrix

```
> A:=matrix([[3,-1,-2],[2,0,-2],[2,-1,-1]]);
```

Is matrix A diagonalizable? First find how many distinct eigenvalues A has

```
> eigenvals(A);
```

Matrix A has two distinct eigenvalues. What are the associated eigenvectors?

```
> eigenvects(A);
```

In this example, the eigenvector corresponding to the eigenvalue 0 is

```
> v1:=vector([1,1,1]);
```

The eigenvectors corresponding to eigenvalue 1 are

```
> v2:=vector([1,2,0]); v3:=vector([0,-2,1]);
```

Is A diagonalizable? Select the nostep mode to see whether matrix A is diagonalizable

```
> diagonalize(A);
```

Although there are only two eigenvalues, there are three independent eigenvectors. Eigenvalue 1 repeats twice and we say its **algebraic** multiplicity is 2. Since there are two eigenvectors associated with this eigenvalue, we say its **geometric** multiplicity is 2. Thus we can construct the matrix whose columns are eigenvectors v_1, v_2, and v_3 of matrix A:

```
> P:=transpose(matrix([v1,v2,v3]));
```

Is P a nonsingular matrix?

```
> det(P);
```

Perform the matrix multiplication $P^{-1}AP$.

```
> P^('-1')*A*P = multiply(inverse(P),multiply(A,P));
```

In this example, matrix A is similar to a diagonal matrix.

What if the algebraic multiplicity of an eigenvalue is not equal to its geometric multiplicity? Can we still diagonalize a matrix A?

EXAMPLE 2.4   Consider the 3×3 matrix

```
> A:=matrix([[1,0,0],[0,2,0],[3,0,1]]);
```

Is A diagonalizable? Use the nostep mode of the diagonalize function

```
> diagonalize(A);
```

Find out the number of independent eigenvectors of matrix A

```
> eigenvects(A);
```

There are only two independent eigenvectors. One is

```
> v1:=vector([0,1,0]);
```

and the other is an eigenvector corresponding to the eigenvalue 1 of algebraic multiplicity 2

```
> v2:=vector([0,0,1]);
```

Can we construct a 3×3 matrix P such that $P^{-1}AP$ is a diagonal matrix? To do so, we need three independent eigenvectors to form the columns of matrix P. We have only two. In this example, matrix A is not diagonalizable. The matrix A is referred to as a **defective matrix**.

An *n* x *n* matrix A is diagonalizable if there exists a nonsingular matrix P such that the product $P^{-1}AP$ is a diagonal matrix.

The algebraic multiplicity of an eigenvalue is the number of times the root of the characteristic polynomial repeats; its geometric multiplicity is the dimension of the eigenspace associated with that eigenvalue.

Let A be an *n* x *n* matrix. If A has n distinct eigenvalues, then A has n linearly independent eigenvectors and A is diagonalizable. The eigenvectors are the columns of the diagonalizing matrix P. The matrix P is unique up to the order of its columns. (Fact 6.3)

Let A be an $n \times n$ matrix. A is diagonalizable if and only if A has n independent eigenvectors. (Fact 6.4)

APPLICATION

Exponential of a Matrix

The differential equation $\frac{dy(t)}{dt} = y'(t) = ay(t)$ is a prototype equation that models problems of growth and decay. For example, if y(t) is the count of bacteria at time t and if the rate of change of bacteria is proportional to the count at time t, then the differential equation models this phenomenon with a being the constant of proportionality. The solution is $y(t) = ce^{at}$ and c is an arbitrary constant that can be determined using the initial condition y(0).

Recall that the exponential function e^{at} is given by the expansion

```
> exp(a*t):=Sum( (a*t)^i/i!, i=0..infinity);
```

Since this function is a solution of many important problems including the problem of exponential growth or decay, a natural question to ask in this setting is there an equivalent expansion where the parameter a is replaced by a matrix A? If so, how do we compute this expression?

This result would be needed if we were to replace the scalar equation by the matrix equation $Y'(t) = AY(t)$, where $Y(t) = [y_1(t), y_2(t), \ldots, y_n(t)]$ and solve the resulting system. The solution of this system is $Y(t) = e^{At}Y(0)$ where $Y(0)$ is the initial condition. To have an analogous exponential function for matrices

```
> exp(A*t):=Sum( (A*t)^i/i!, i=0..infinity);
```

we need to compute the powers of the matrix A for different type of matrices.

DIAGONAL MATRIX

Let A be the 2×2 identity matrix

```
> A:=diag(1,1);
```

Since all powers of A are equal to A

```
> exp(A*t):=evalm(A)*Sum( (t)^i/i!, i=0..infinity);
```

This is equivalent to

```
> exp(A*t):=evalm((A)*sum( t^i/i!, i=0..infinity));
```

Now let A be the 2×2 diagonal matrix

```
> A:=diag(2,3);
```

Since A is a diagonal matrix, its powers are easy to compute. In particular, for matrix A

```
> A^2=evalm(A^2); A^3=evalm(A^3); A^4=evalm(A^4);
```

In general,

```
> A^i = matrix([[2^i,0],[0,3^i]]);
```

Therefore,

```
> exp(A*t):=Sum(matrix([[2^i,0],[0,3^i]])*t^i/i!,
i=0..infinity);
```

That is,

```
> exp(A*t):=Sum(matrix([[(2*t)^i/i!,0],[0,(3*t)^i/i!]]),
i =0..infinity);
```

or

```
> exp(A*t):= matrix([[Sum((2*t)^i/i!,i=0..infinity),0],
[0,Sum((3*t)^i/i!,i=0..infinity)]]);
```

When the sums are evaluated, we get the matrix representation of e^{At}

```
> exp(A*t):= matrix([[sum((2*t)^i/i!,i=0..infinity),0],
[0,sum((3*t)^i/i!,i=0..infinity)]]);
```

DIAGONALIZABLE MATRIX

Now consider the 2×2 matrix

```
> A:=matrix([[3,4],[3,2]]);
```

The eigenvalues of matrix A are 6 and -1 and the corresponding eigenvectors are

```
> v1:=vector([4,3]); v2:=vector([1,-1]);
```

Thus matrix A is similar to a diagonal matrix with

```
> P:=matrix([[4,1],[3,-1]]):D1:=diag(6,-1):
evalm(A)=evalm(P)*evalm(D1)*evalm(inverse(P));
```

Let us use the fact that A is diagonalizable to compute various powers of A

```
> A^2=evalm(A^2);
P*D1^2*P^('-1')=multiply(P,multiply(D1^2,inverse(P)));
> A^3=evalm(A^3);
P*D1^3*P^('-1')=multiply(P,multiply(D1^3,inverse(P)));
>  A^4=evalm(A^4);
P*D1^4*P^('-1')=multiply(P,multiply(D1^4,inverse(P)));
```

Then

```
> A^n=P*D1^n*P^('-1');
```

Furthermore, the exponential of the matrix D_1 is

```
> exp(D1*t):=diag(exp(6*t),exp(-t));
```

Thus,

```
> exp(A*t):=evalm(P&*exp(D1*t)&*inverse(P));
```

SUMMARY

To get the exponential of a diagonalizable matrix A proceed as follows:

1. Compute the eigenvalues and eigenvectors of the given matrix.
2. Construct the matrix P whose columns are the eigenvectors and form the product $A = PD_1P^{-1}$, where D_1 is the diagonal matrix whose diagonal entries are the eigenvalues of A.
3. Construct the exponential matrix $e^{D_1 t}$.
4. The exponential of the given matrix is $Pe^{D_1 t}P^{-1}$.

INITIAL VALUE PROBLEM

Solve the initial value problem

```
> A:=matrix([[3,2],[0,4]]):
Y:=matrix([[y1],[y2]]):
> evalm((d/dt)*Y) = evalm(A)*evalm(Y);
```

with initial data Y(0)

```
> Y0:=matrix([[1],[2]]);
```

1. Compute the eigenvalues of A (use the nostep mode)

```
> eigenvals(A);
```

2. Compute the eigenvectors of matrix A (use the nostep mode)

```
> eigenvects(A);
```

An eigenvector corresponding to the eigenvalue 3 is the vector [1, 0], and an eigenvector corresponding to the eigenvalue 4 is the vector [2, 1].

3. Construct the matrix P and the diagonal matrix D_1

```
> P:=matrix([[1,2],[0,1]]); D1:=diag(3,4);
```

4. Construct the exponential of D_1

```
> exp(D1*t):=diag(exp(3*t),exp(4*t));
```

The exponential matrix of A is

```
> exp(A*t):=evalm(P&*exp(D1*t)&*inverse(P));
```

Therefore, the solution to the initial value problem is

```
> Y:=evalm(exp(A*t)&*Y0);
```

EXERCISES

In the following exercises you may need to use the automated functions `eigenvals`, `eigenvects`, and `diagonalize`, and the Maple commands `solve`, `matrix`, `vector`, `gausselim`, `rref`, and `charpoly`.

1. Can you find a diagonal matrix D_1 similar to the matrix

```
> A:=matrix([[1,0,0,0],[0,1,0,0],[0,0,-2,-1],[0,0,-1,-2]]);
```

Use the interactive mode of the function `diagonalize`.
Deduce from the result the tenth power of A.

2. Let T be the linear transformation

```
> T:=(x1,x2,x3)->(x1+x2,x2+x3,0);
```

Is there a basis B for $\mathbb{R}^3$ for which the matrix representing T relative to basis B is diagonal? If so, find basis B and the diagonal matrix representation.

3. Let $A = (a_{ij})$ be an $n \times n$ upper triangular matrix with distinct diagonal elements.

   a. Is A similar to a diagonal matrix? *Hint*: Experiment with several examples and then make a general statement.
   b. Repeat part (a) letting A be a lower triangular matrix with distinct diagonal elements.

4. Solve the initial value problem

```
> dY/dt = A*Y;
```

with initial data $Y(0)$ and matrix A being

```
> Y0:=matrix([[1],[1],[1]]);
    A:=matrix([[1,1,1],[0,0,1],[0,0,-1]]);
    Y:=matrix([[y1],[y2],[y3]]);
```

Hermitian Matrices

We have dealt so far with matrices whose entries are real numbers. What about matrices whose entries are complex numbers? Recall that a complex number is a number of the form $a + bi$ where a and b are real numbers and $i = \sqrt{-1}$.

Initialize the packages

```
> with(linalg):with(lineign);
```

EXAMPLES OF A HERMITIAN MATRIX

EXAMPLE 3.1   Let A be the matrix

```
> A:=matrix([[1,2,-I],[2,3,1+2*I],[I,1-2*I,0]]);
```

Is A equal to its transpose?

```
> A^T=transpose(A);
```

What if we take the transpose of the matrix whose entries are the complex conjugates of the entries of the matrix A

```
> (conjugateA)^T=transpose(matrix([[1,2,I],[2,3,1-2*I],
  [-I,1+2*I,0]]));
```

Matrix A is equal to its conjugate. A matrix such as matrix A is called a **Hermitian matrix.** These matrices are the complex analogues of the symmetric matrices whose entries are real numbers.

EXAMPLE 3.2   Is the following matrix hermitian?

```
> A:=matrix([[2,2-I],[2+I,4]]);
```

Take the conjugate of each entry of matrix A

```
> conjugate(A):=matrix([[2,2+I],[2-I,4]]);
```

and the transpose of conjugate(A) is

```
> transpose(conjugate(A));
```

Since this matrix is equal to matrix A, A is hermitian.

LEARNING THE PROCESS

Use the demonstration mode of the function `hermitian` to check whether the following matrix is hermitian.

EXAMPLE 3.3   Is the following matrix hermitian?

```
> A:=matrix([[I,2-I],[2+I,4]]);
> hermitian(A);
```

Repeat using your own matrices until you have mastered the process.

SOME PROPERTIES OF HERMITIAN MATRICES

EXAMPLE 3.4   Given the 2×2 hermitian matrix A

```
> A:=matrix([[2,2-I],[2+I,4]]);
```

what are the eigenvalues of the hermitian matrix (use the nostep mode)A?

```
> eigenvals(A);
```

The eigenvalues of matrix A are real numbers. Is this always the case?

What are the eigenvectors of the hermitian matrix A (use the nostep mode)?

```
> eigenvects(A);
```

The two eigenvectors ev_1 and ev_2 are

```
> ev1:=vector([2-I,1-sqrt(6)]);
ev2:=vector([-(1-sqrt(6)),2+I]);
```

What is the inner product of ev_1 and ev_2? To define the inner product for vectors with complex entries, take the inner product of the first vector with the conjugate of the second vector. The conjugate of ev_2 is

```
> cev2:=vector([-(1-sqrt(6)),2-I]);
```

and the innerproduct of the eigenvectors ev_1 and ev_2 is

```
> innerprod(ev1,cev2);
```

The eigenvectors corresponding to distinct eigenvalues are orthogonal. Is this always the case?

EXAMPLE 3.5   Consider the matrix

```
> A:=matrix([[1,0,-I],[0,1,1+2*I],[I,1-2*I,0]]);
```

Determine the eigenvalues of A (use the nostep mode)

```
> eigenvals(A);
```

The eigenvalues of matrix A are real numbers. Is this always the case? What are the eigenvectors of the hermitian matrix A (use nostep mode)?

```
> eigenvects(A);
```

The eigenvectors are

```
> ev1:=vector([-1/2*I,1/2+I,1]);
ev2:=vector([1/3*I,-1/3-2/3*I,1]);
ev3:=vector([2+I,1,0]);
```

Before computing the inner products of the eigenvectors, take the conjugate of the vectors

```
> cev1:=vector([1/2*I,1/2-I,1]);
cev2:=vector([-1/3*I,-1/3+2/3*I,1]);
cev3:=vector([2-I,1,0]);
```

Take the following inner products:

```
> '< ev1 , ev2 >' =innerprod(ev1,cev2);
> '< ev1 , ev3 >'=innerprod(ev1,cev3);
> '< ev2 , ev3 >'=innerprod(ev2,cev3);
```

Again, the eigenvectors of the hermitian matrix corresponding to the distinct eigenvalues are orthogonal.

The eigenvalues of a hermitian matrix are real. The eigenvectors of a hermitian matrix are orthogonal. (Fact 6.5)

EXERCISES

In the following exercises you may need to use the automated functions eigenvals, eigenvects, diagonalize, and hermitian and the Maple commands solve, matrix, vector, gausselim, rref, and charpoly.

1. Let A and B be two hermitian matrices

```
> A:=matrix([[1,-I,0],[I,1,2],[0,2,1]]);
    B:=matrix([[0,1+I,2],[1-I,2,I],[2,-I,3]]);
```

  a. Is the sum of A and B hermitian?
  b. Is the scalar product $c * A$ hermitian (1) if c is real? (2) if c is complex?
  c. Is the product of A and B hermitian?
  d. Are the powers of A hermitian?

2. Consider the matrix

```
> A:=matrix([[1,-I,0],[I,2,0],[0,0,1]]);
```

  a. Find the eigenvalues and eigenvectors of matrix A.

  b. Construct an orthogonal matrix P such that $P^{-1}AP$ is a diagonal matrix.

3. Find all 2×2 and 3×3 hermitian matrices. List a few examples of each type.

Unitary Matrices

The class of unitary matrices is the complex analog of the class of orthogonal matrices. A unitary matrix U is a matrix with complex entries whose columns form an orthonormal set. Given a hermitian matrix A, can we find a unitary matrix U such that $U^{-1}AU = D$ where D is a diagonal matrix?

Initialize the packages

```
> with(linalg):with(lineign);
```

EXAMPLE OF A UNITARY MATRIX

EXAMPLE 4.1   Consider the hermitian matrix

```
> A:=matrix([[2,1-I],[1+I,1]]);
```

Find the eigenvalues of the matrix A (use the nostep mode)

```
> eigenvals(A);
```

The eigenvalues of the matrix A are 0 and 3.
Construct the eigenvectors:

```
> eigenvects(A);
```

The eigenvectors corresponding to the eigenvalues 0 and 3 are

```
> ev1:=vector([1,-1-I]); ev2:=vector([1,1/2+1/2*I]);
```

Now compute the conjugates of ev_1 and ev_2

```
> cev1:=vector([1,-1+I]); ev2:=vector([1,1/2-1/2*I]);
```

Are the eigenvectors orthogonal?

```
> '< ev1 , ev2 >' = innerprod(ev1,cev2);
  '< ev2 , ev1 >'= innerprod(ev2,cev1);
```

Normalize ev_1 and ev_2

```
> u1:=evalm((1/sqrt(innerprod(ev1,cev1)))*ev1);
  u2:=evalm((1/sqrt(innerprod(ev2,cev2)))*ev2);
```

The matrix U

```
> U:=transpose(matrix([u1,u2]));
```

is an example of a unitary matrix. Its columns form an orthonormal set.

SOME PROPERTIES OF UNITARY MATRICES

What is the inverse of the unitary matrix in Example 4.1?

```
> U^(' -1') = inverse(U);
```

What is the relation between the entries of U and its inverse? The inverse of a unitary matrix U can be obtained by taking the transpose of the conjugate of U.

What is the product $U^{-1}AU$?

```
> U^(' -1')*A*U=evalm(inverse(U)&*A&*U);
```

On simplification

```
> evalf(evalm(inverse(U)&*A&*U));
```

the product yields a diagonal matrix with diagonal elements being the eigenvalues of the hermitian A.

If a hermitian matrix A has distinct evalues, then there is a unitary matrix U that diagonalizes A. (Fact 6.6)

EXERCISES

In the following exercises you may need to use the automated functions `eigenvals`, `eigenvects`, `diagonalize`, and `hermitian` and the Maple commands `solve`, `matrix`, `vector`, `gausselim`, `rref`, and `charpoly`.

1. Let U be a unitary matrix

```
> U:=matrix([[0,1/sqrt(2),1/sqrt(2)],[-1, 0, 0],
   [0,-1/sqrt(2),1/sqrt(2)]]);
```

  a. Are the powers of U unitary?
  b. Show that the absolute value of the eigenvalues of U is always equal to 1.
  c. Given any two vectors v_1 and v_2 in $\mathbb{R}^3$, show that, with respect to the standard inner product of $\mathbb{R}^3$, the inner product of Uv_1 and Uv_2 is equal to the inner product of v_1 and v_2.

2. Given the matrix

```
> A:=matrix([[0,2,-1],[2,3,-2],[-1,-2,0]]);
```

find an orthogonal matrix U that diagonalizes A.

3. Find values for the parameters a_1, a_2 and a_3 so that the matrix U is unitary

```
> U:=matrix([[a1,sqrt(6)/3,-sqrt(3)/3],
    [a2,sqrt(6)/6,sqrt(3)/3], [a3,sqrt(6)/6,sqrt(3)/3]]);
```

Quadratic Forms and Positive Definite Matrices

The concepts of eigenvalues and eigenvectors lead naturally to the study of quadratic forms and positive definite matrices.

Initialize the packages

```
> with(linalg):with(lineign);
```

QUADRATIC FORMS

A quadratic form in two variables x and y is

```
> a*x^2+b*x*y+c*y^2+d*x+e*y+f=0;
```

This quadratic form represents, geometrically, a conic. Recall that by applying the process of completing the squares to a quadratic equation, we can put it in a standard form from which we can identify the conic and its characteristics (foci, center, and so axes). The conic can be identified based on the coefficients a, b, and c.

- When a = c, the equation represents a circle.
- When a and c have same sign, the equation represents an ellipse.
- When a and c have opposite signs, the equation represents a hyperbola.
- When a = 0 or c = 0, the equation represents a parabola.

Let us write the quadratic form in matrix form. Let X and A be the two matrices

```
> X:=matrix([[x],[y]]); A:=matrix([[a,b],[b,c]]);
```

The quantity

```
> X^T*A*X=multiply(transpose(X),multiply(A,X));
```

is called the **quadratic form** associated with the quadratic equation. Note that matrix A is symmetric.

APPLICATIONS

EXAMPLE 5.1   Consider the conic

```
> 3*x^2+2*x*y+3*y^2-8=0;
```

Its graph is

```
> with(plots):
> implicitplot(3*x^2+2*x*y+3*y^2-8=0,x=-5..5,y=-5..5);
```

In this example, it is not easy to determine some of the characteristics of the conic, such as its center, foci, and major and minor axes, because it is not in a standard form. The term 2xy is creating a rotation of axes. Set X and A to be

```
> X:=matrix([[x],[y]]);A:=matrix([[3,1],[1,3]]);
```

The equation can be written

```
> X^T*A*X=8;
```

That is,

```
> multiply(transpose(X),multiply(A,X))=8;
```

What are the characteristics of this ellipse? To determine the characteristics, put this equation in a standard form.

Let us see if the eigenvalues and eigenvectors of the matrix A can help in this process.

1. Determine the eigenvalues of the symmetric matrix A (use the nostep mode)

```
> eigenvals(A);
```

Notice that matrix A has two distinct real eigenvalues, 4 and 2.

2. Determine the eigenvectors of matrix A (use the nostep mode).

```
> eigenvects(A);
```

The eigenvectors corresponding to the eigenvalues 2 and 4 are, respectively,

```
> ev1:=vector([-1,1]);ev2:=vector([1,1]);
```

Therefore matrix A is diagonalizable.

3. Normalize the orthogonal eigenvectors ev_1 and ev_2

```
> a1:=sqrt(innerprod(ev1,ev1)):
a2:=sqrt(innerprod(ev2,ev2)):
> u1:=evalm(1/a1*ev1);u2:=evalm(1/a2*ev2);
```

Construct the orthogonal matrix (rotation matrix) Q whose columns are the vectors u_1 and u_2

```
> Q:=transpose(matrix([u1,u2]));
```

4. Matrix A is similar to a diagonal matrix D_1 since

```
> Q^T*A*Q=D1;
```

where

```
> D1:=multiply(transpose(Q), multiply(A,Q));
```

5. Introduce a new coordinate system X_1

```
> X1:=matrix([[x1],[y1]]);
```

The relation between the original coordinate system X and the new coordinate system X_1 is

```
> evalm(X)=multiply(Q,X1);
```

With respect to the new rotated axes, the original quadratic expression

```
> X^T*A*X=multiply(transpose(X),multiply(A,X));
```

is

```
> multiply(transpose(X1), multiply(D1,X1))=8;
```

This can be simplified to the standard form of the conic section

```
> 1/2*x1^2+1/4*y1^2=1;
```

Thus, the characteristics of the ellipse with respect to the new coordinate system are easily determined.

Minimizing Functions of Several Variables

Another application is determining the **global extrema** of a function of several variables. Let us assume that f(x, y) is a function of two variables

```
> f:=(x,y)->a*x^2+2*b*x*y +c*y^2;
```

To determine the extrema of this function, first find the derivative of f(x, y) with respect to each variable:

```
> delta(f)/delta(x)=diff(f(x,y),x);
delta(f)/delta(y)=diff(f(x,y),y);
```

By setting each equation to 0 and solving the resulting homogeneous system, we get the critical point (0, 0). It will be the only critical point, provided that the coefficient matrix is nonsingular. Put the function in a quadratic form:

```
> X:=matrix([[x],[y]]); A:=matrix([[a,b],[b,c]]);
> f:=X->multiply(transpose(X),multiply(A,X));
> 'f(X)'=f(X);
```

Is the critical point (0, 0) a minimum, maximum, or saddle point? The answer depends upon one of the following conditions.

- The critical point is a global minimum provided that the quadratic form

```
> X^T*A*X=multiply(transpose(X),multiply(A,X));
```

is strictly positive for all X different from zero. In this case symmetric matrix A is said to be **positive definite**. This is equivalent to saying that the eigenvalues of A are positive.

- The critical point is a global maximum provided that the quadratic form

```
> X^T*A*X=multiply(transpose(X),multiply(A,X));
```

is strictly negative for all X different from zero. In this case symmetric matrix A is said to be **negative definite**. This is equivalent to saying that the eigenvalues of A are negative.

- The critical point is a saddle point provided that the quadratic form

```
> X^T*A*X=multiply(transpose(X),multiply(A,X));
```

changes sign. In this case the quadratic form is **indefinite**.

EXAMPLE 5.2 Classify the critical points of the function

```
> f:=(x,y)->2*x^2-4*x*y+5*y^2;
```

Its quadratic form representation is

```
> X:=matrix([[x],[y]]); A:=matrix([[2,-2],[-2,5]]);
> f:=X->multiply(transpose(X),multiply(A,X));
> 'f(X)'=f(X);
```

The eigenvalues of the matrix A are

```
> eigenvals(A);
```

Both eigenvalues are positive. What does this tell about the quadratic form? For the eigenvalue 1, the eigenvector x satisfies Ax = x. Compute the quadratic form:

```
> X^T*A*X=multiply(transpose(X),X);
```

For the eigenvalue 6, the eigenvector X satisfies Ax = 6x. Compute the quadratic form

```
> X^T*A*X=multiply(transpose(X),evalm(6*X));
```

In either case, the quadratic form is positive definite (note that both eigenvalues are positive). Therefore, the origin is a global minimum.

EXAMPLE 5.3 Consider the function

```
> f:=(x,y)->-x^2+2*x*y-5*y^2;
```

Its quadratic form representation is

```
> X:=matrix([[x],[y]]); A:=matrix([[-1,1],[1,-5]]);
> f:=X->multiply(transpose(X),multiply(A,X));
> 'f(X)'=f(X);
```

Compute the eigenvalues of matrix A:

```
> eigenvals(A);
```

Both eigenvalues are negative. What does this tell us about the quadratic form?

For the eigenvalue $-3 + \sqrt{5}$, $Ax = (-3 + \sqrt{5})x$ for eigenvector x. Compute the quadratic form

```
> X^T*A*X=multiply(transpose(X),evalm((-3+sqrt(5))*X));
```

The quadratic form is negative definite.
For the eigenvalue $-3 - \sqrt{5}$, $Ax = (-3 - \sqrt{5})x$ for an eigenvector x. Compute the quadratic form

```
> X^T*A*X=multiply(transpose(X),evalm((-3-sqrt(5))*X));
```

The quadratic form is negative definite. Therefore, in this example the origin is a global maximum.

A quadratic form is positive definite if and only if all eigenvalues are positive. In this case the critical point is a global minimum.

A quadratic form is negative definite if and only if all eigenvalues are negative. In this case the critical point is a global maximum.

EXERCISES

In the following exercises you may need to use the automated functions `eigenvals`, `eigenvects`, `diagonalize`, and `hermitian` and the Maple commands `solve`, `matrix`, `vector`, `gausselim`, `rref`, and `charpoly`.

1. To identify the characteristics of the quadratic equation $3x^2 + 8xy + 3y^2 + 28 = 0$, reduce the equation into a standard form.

2. Consider the matrix

```
> A:=matrix([[a,b],[b,c]]);
```

What type of conic section will the equation

```
> a*x^2+2*b*x*y+c*y^2=1;
```

represent if the product of the eigenvalues of A is negative? Explain.

3. Find the values of c for which the matrix

```
> A:=matrix([[c,1,1],[1,c,1],[1,1,c]]);
```

is a positive definite matrix.

4. Given the function

```
> f:=(x,y) ->-x^2+4*x*y-3*y^2;
```

is the origin global maximum, global minimum, or a saddle point?

Eigenvalues and Discrete Systems

Purpose

The objective of this lab is to enforce the concepts of eigenvalues and eigenvectors and apply them to analyze discrete dynamical systems.

Automated Linalg functions

In this lab you will use the automated functions `eigenvals`, `eigenvects`, `diagonalize`, and `hermitian`. To get help with a function, type at the Maple prompt sign >function name; for example, >?eigenvals;

Instructions

1. To execute a statement, move the cursor to the line using the mouse or the keyboard and press the enter key.
2. While executing a function, create Maple input regions if needed; otherwise the output will not appear in the desired place.
3. Execute the following commands to load the packages.

```
> with(linalg):with(lineign);
```

TASK 1

The purpose of this task is to elaborate on the concepts of eigenvalues and eigenvectors. Suppose we are given the matrix

```
> A:=matrix([[45,-21,-63,-12,21],[28,-12,-42,-7,14],
[31,-15,-43,-9,15],[28,-14,-42,-5,14],
[51,-25,-75,-14,27]]);
```

Activity 1

Find the eigenvalues of matrix A (use the interactive mode of `eigenvals`)

```
> eigenvals(A);
```

What is the algebraic multiplicity of each eigenvalue?

Activity 2

Find the eigenvectors (use the interactive mode of `eigenvects`)

```
> eigenvects(A);
```

Enter the eigenvectors as v_1, v_2, v_3, v_4, and v_5. What is the geometric multiplicity of each eigenvalue?

Activity 3

Based on Activity 2, is matrix A similar to a diagonal matrix? Explain your answer.

Activity 4

Construct a matrix P whose columns are the eigenvectors of Activity 3. Is matrix P unique? Explain.

Activity 5

Show that A is similar to a diagonal matrix D.

Activity 6

Show that the trace of matrix A is equal to the sum of the eigenvalues.

Activity 7

Compute the products PD^iP^{-1} for $i = 2, 3, 4$. How do these products compare with A^i for $i = 2, 3, 4$?

Activity 8

For the given matrix A and the vector

```
> Y:=matrix([[y1],[y2],[y3],[y4],[y5]]):
```

solve the inital value problem

```
> evalm((d/dt)*Y)=evalm(A)*evalm(Y);
```

subject to the initial condition Y_0

```
>   Y0:=matrix([[1],[0],[-1],[4],[3]]);
```

Use the results obtained in Activity 7 to write the solution. *Hint*: Construct e^{At}.

TASK 2

The purpose of this task is to use eigenvalues and eigenvectors to analyze the behavior of discrete dynamical systems. Consider the system defined by the difference equation $x_{k+1} = Ax_k$ where

```
> A:=matrix([[0.9,0.04],[0.1,0.96]]);
```

and the initial value x_0 of x is

```
> x(0):=matrix([[.3],[.5]]);
```

Activity 1

Compute the eigenvalues of matrix A

```
> eigenvals(A);
```

Compute the eigenvectors for each eigenvalue

```
> eigenvects(A);
```

Activity 2

Express the initial condition x_0 as a linear combination of the eigenvectors of Activity 1.

Activity 3

Compute the first iterates $x_1, x_2, x_3, \ldots$ using the result of Activity 2.

Activity 4

Deduce an expression for x_k. Determine the behaviour of the solution as k approaches infinity.

EXTRA LAB PROBLEMS

1. Consider the matrix

```
> A:=matrix([[0,1,0],[1,1,1],[0,1,0]]);
```

   Can you propose a general relation between eigenvalues and eigenvectors of matrix A and its powers? Verify your answer.

2. Consider the matrix

```
> A:=matrix([[1,1,1],[1,1,1],[1,1,1]]);
```

Show, by using only matrix multiplication, that the vector

```
> v:=vector([1,1,1]);
```

is an eigenvector with an eigenvalue 3.

3. Can you propose a similar statement for an $n \times n$ matrix with all entries equal to 1? Is matrix A diagonalizable? Explain your answer.

4. Experiment with several matrices of different sizes but with entries all equal to 1 to answer the questions posed in problems 2 and 5.

Quadratic Forms

Purpose

The objective of this lab is to analyze the geometric significance of eigenvectors and eigenvalues and their properties and applications, in particular, quadratic forms and their applications.

Automated Linalg functions

In this lab you will use the automated functions `eigenvals`, `eigenvects`, `diagonalize`, and `hermitian`. To get help with a function, type at the Maple prompt sign >function name; for example, >?eigenvals;

Instructions

1. To execute a statement, move the cursor to the line using the mouse or the keyboard and press the enter key.
2. While executing a function, create Maple input regions if needed; otherwise the output will not appear in the desired place.
3. Execute the following commands to load the packages.

    ```
    > with(linalg):with(lineign);
    ```

TASK 1

The purpose of this task is to review the notion of hermitian matrices. Consider the matrix

```
> A:=matrix([[3,1+I,I],[1-I,1,3],[-I,3,1]]);
```

Activity 1

Is matrix A a hermitian matrix?

Activity 2

Find the eigenvalues of matrix A.

Activity 3

Find the eigenvectors of matrix A.

Activity 4

Show that the eigenvectors of matrix A are orthogonal.

Activity 5

Construct a matrix U that diagonalizes matrix A.

TASK 2

The purpose of this task is to enforce the study of quadratic forms and their applications. Consider the quadratic equation

```
> eq1:=2*x^2 +2*x*y+2*y^2=9;
```

Activity 1

Execute the following Maple command to graph the quadratic equation

```
> with(plots):
> implicitplot(eq1,x=-10...10,y=-10..10,color=red);
```

From the plot, can you identify the characteristics of this conic (center, foci, and major and minor axes)?

Activity 2

Express eq_1 in a quadratic form. That is, define the matrix A and the vector X so that the product $X^T A X$ is eq_1.

Activity 3

Find the eigenvalues and eigenvectors of matrix A.

Activity 4

Are the eigenvectors of matrix A orthogonal?

Activity 5

Define the transformation from the original xy-coordinate system to a new $x_1 y_1$-coordinate system. To put the original quadratic form into a standard form, substitute the transformed coordinates into the original form.

Activity 6

Use the standard form to identify the conic section (center, foci, and major and minor axes) with respect to the new coordinate axes.

EXTRA LAB PROBLEM

Define the function

```
> f:=(x,y) ->x^2+4*x*y+4*y^2-12;
```

Find the minimum of the function $f(x, y)$ subject to the condition

```
> x^2+y^2=1;
```

a. Write the quadratic form corresponding to the first three terms in f. That is, identify the matrix A and the vector X for which $X^T A X$ is equal to these terms.
b. Compute the eigenvalues and eigenvectors of matrix A.
c. Normalize the eigenvectors of matrix A.
d. Compute the value of $f(x, y)$ at each of the eigenvectors. Which eigenvector yields the minimum value of $f(x, y)$?
e. Can you propose a general statement from your observations?

Ecological Model

Purpose

The purpose of this application is to analyze an ecological model and predict its behavior using the notions of eigenvalues and eigenvectors.

Initialize the packages.

```
> with(linalg):with(lineign);
```

DESCRIPTION OF AN ECOLOGICAL MODEL

Consider a habitat in which sheep (S) and foxes (F) compete for survival. The number of sheep (S) grows by 20% a year in the absence of foxes and decreases by 12% of the total fox population. In the absence of sheep, the number of foxes (F) decreases by 22% and increases by 30% of the total sheep population.

Activity 1

Write the mathematical equations that describe the population of sheep (S_1) and foxes (F_1) after one year. Represent this model in matrix form.

Activity 2

We want to analyze the behaviour of this population as it evolves over time. Assume that the initial population is $S_0 = 1000$ and $F_0 = 100$. What is the population of sheep and foxes after one year? Use matrix multiplication to answer this question. Predict the size of each population after 2, 3, 10, and 20 years? Do you think both sheep and fox populations will continue to grow?

Activity 3

Write the coefficient matrix of the model in Activity 1. Compute its eigenvalues and eigenvectors.

Activity 4

Express the initial population size as a linear combination of the eigenvectors determined in Activity 3.

Activity 5

Can you deduce the size of each population after 100 years? In general, after n years?

Activity 6

Does the size of each population reach a steady state? How does your answer relate to the eigenvalues of the matrix?

Activity 7

Instead of 22%, consider the decrease in fox population to be in the range of 14 –22%. Within this range, which percentage leads to an extinction of the population?

Shopping Strategy

Purpose

The purpose of this application is to decide on a shopping strategy and show how the notions of eigenvalues and eigenvectors can be utilized in the mathematical modeling and the analysis of this problem.

Initialize the packages

```
> with(linalg):with(lineign);
```

DESCRIPTION OF A SHOPPING STRATEGY MODEL

Each week a shopper buys a certain product and chooses from one of three brands. The decision about which brand to buy is based on which brand was bought last week. If the shopper bought

- brand 1 last week, then he will buy brand 1 with probability 0.8, brand 2 with probability 0.1, and brand 3 with probability 0.1.
- brand 2 last week, then he will buy brand 1 with probability 0.1, brand 2 with probability 0.8, and brand 3 with probability 0.1.
- brand 3 last week, then he will buy brand 1 with probability 0.2, brand 2 with probability 0.2, and brand 3 with probability 0.6.

The following table summarizes the probabilities of buying a brand this week based on what was bought the previous week.

	Brand1	Brand 2	Brand3
Brand1	0.8	0.1	0.1
Brand2	0.1	0.8	0.1
Brand3	0.2	0.2	0.3

The probabilities represent transition probabilities. Let B_n be the brand bought in week n. Thus $B_n = 1$ represents that brand 1 was bought in week n. Let $pr(B_n = 1)$ denote the probability that brand 1 was bought in week n. Therefore, the probability vector for week n is

```
> p(n):=matrix([[pr(B(n)=1)],[pr(B(n)=2)],[pr(B(n)=3)]]);
```

Activity 1

Describe the probability matrix A of buying the different brands in two successive weeks (for example, week 0 and week 1).

Activity 2

If the shopper buys brand 1 in week 0, the probability vector in week 0 is

```
> p(0):=matrix([[1],[0],[0]]);
```

What are the probability vectors of each brand in week 1? week 2? week 3? week 4?

Activity 3

Describe the probability matrix of buying the different brands in week $n+1$ and week n.

Activity 4

Show that the matrix A is similar to a diagonal matrix. *Hint*: Compute the eigenvectors of the matrix A.

Activity 5

Use the result of Activity 4 to compute A^n. Then deduce the limit of A^n as $n \to \infty$.

Activity 6

Does the limit p_n exist as n increases indefinitely? Does the limit depend on which brand was bought in week 0? Explain your assertion carefully.

Activity 7

Is there a relation between the results obtained in Activity 6 and Activity 5? Explain your answer.

Concentration of Carbon Dioxide in the Blood

Purpose

In this application we study and analyze the concentration of carbon dioxide in the blood. The notions of eigenvalues and eigenvectors are used to analyze this model.

Initialize the packages

```
> with(linalg):with(lineign);
```

DESCRIPTION OF CO_2 CONCENTRATION MODEL

The process of producing and losing carbon dioxide (CO_2) in the blood is a continuous nonlinear process. However, we shall assume a **discretization** of this process by assuming that

- breathing takes place at constant time intervals $t, t + a, t + 2a, t + 3a, \ldots$, and the ventilation volume $V_n = V_{t+na}$ is controlled by the CO_2 concentration in the blood level $C_{n-1} = C_{t+(n-1)*a}$ in the previous time interval.
- the concentration of CO_2 at time $t + (n + 1)a$, C_{n+1}, is a function of C_n minus the amount of CO_2 lost plus the amount produced due to metabolism. That is, $C_{n+1} =$ (amount of CO_2 previously in the blood) $-$ (amount of CO_2 lost) $+$ (constant production due to metabolism).

We shall also consider a **linearization** for this process by assuming that

- the amount of CO_2 lost is proportional to the ventilation volume V_n with constant of proportionality c_1 and does not depend upon C_n.
- the ventilation volume V_{n+1} is directly proportional to C_n with a constant of proportionality c_2.
- the CO_2 produced due to metabolism is held at a constant value k.

Activity 1

Write the mathematical equations that describe this process.

Activity 2

Write the matrix formulation for this nonhomogeneous system of linear equations as $X_{n+1} = AX_n + K$, where $X_n = [C_n, V_n]$.

Activity 3

Derive a general formula that will yield C_n, the CO_2 level in the blood, and V_n, the ventilation volume, in terms of the an initial amount C_0 and V_0.

Activity 4

Obtain the eigenvalues of matrix A. Write the solution in terms of the eigenvalues of A.

Activity 5

Assuming that $4c_1c_2 < 1$, will the level of CO_2 reach a steady state regardless of the initial conditions? If so, what is this steady state? What is the corresponding steady state of the ventilation volume?

Activity 6

Assuming that $4c_1c_2 > 1$, will the CO_2 level in the blood reach a steady state or will it oscillate?

Activity 7

If in Activity 6 the quantity c_1c_2 is very large, what happens to the magnitude of the oscillations? What is the frequency of the oscillations?

Activity 8

Choose different values of c_1, c_2, and k and some initial values for C_0 and V_0 and repeat the Activities 2–7 to confirm your assertions.

Red Blood Cell Production

Purpose

In this application we consider the production of red blood cells, using eigenvalues and eigenvectors to analyze the model.

Initialize the packages

```
> with(linalg):with(lineign);
```

DESCRIPTION OF A RED BLOOD CELL PRODUCTION MODEL

This project is a discrete model for the continuous and nonlinear process of the production of red blood cells (RBCs) in the blood. The process of production may be described as follows:

1. A fraction of the RBCs is being daily filtered out and destroyed by the spleen.
2. The bone marrow produces a number of RBCs proportional to the number lost on the previous day.
3. The number of red blood cells should be maintained at some fixed level.

Use the following symbols:

- C_n: number of RBCs in circulation on day n
- P_n: number of RBCs produced by bone marrow on day n
- f: fraction of RBCs destroyed by the spleen
- c: production constant

Activity 1

Write the mathematical equations that represent the numbers C_n and P_n on day n.

Activity 2

Write the matrix equation for the system obtained in Activity 1.

Activity 3

Determine the eigenvalues of the coefficient matrix of Activity 2.

Activity 4

What conditions must be imposed on the eigenvalues to keep the total number of RBCs in circulation, C_n, roughly the same?

Activity 5

Based on Activity 4, what are the implications on the parameter c?

Activity 6

Using the results of Activities 4 and 5, study the behavior C_n.

Unit One   Systems of Linear Equations

SOLVABILITY OF HOMOGENEOUS SYSTEM OF LINEAR EQUATIONS (LESSON 1.2)

Fact 1.1

A homogeneous system of linear equations is always consistent. The solution set is either a singleton or infinite.

Proof

Since the trivial (zero) solution is a solution for any homogeneous system of linear equations, it follows that such a system is always consistent. Let S denote the solution set. We have three cases to consider.

1. The number of equations (n) is less than the number of unknowns (m). In this case, the system will have at least m − n free variables. Thus the system has infinitely many solutions and the solution set S is infinite.

2. The number of equations (n) is equal to the number of unknowns (m). If, on applying elementary row operations to the coefficient matrix of the system, the resulting rows are all nonzero, then the only solution is the zero solution and the solution set is a singleton. However, if at least one row reduces to a zero row, then the system has at least one free variable. The system will have infinitely many solutions and the solution set in infinite.

3. The number of equations (n) is greater than the number of unknowns (m). On applying elementary row operations to the coefficient matrix of the system, we either get n rows that are nonzero or p rows that are nonzero with p < m. In this case, the problem reduces to case 1 or case 2.

SOLVABILITY OF NONHOMOGENEOUS SYSTEM OF LINEAR EQUATIONS (LESSON 1.2)

Fact 1.2

Every system of linear equations $Ax = b$ has either no solution, exactly one solution, or infinitely many solutions.

Proof

If the system has no solution, we are done. Assume that system $Ax = b$ has a solution x. We will show that if the system has more than one solution then it must have infinitely many solutions. Let x_1 and x_2 be two such solutions. Then $Ax_1 = b$ and $Ax_2 = b$ imply that $A(x_1 - x_2) = 0$. Set $x = x_1 - x_2$. Then $y = x_1 + rx$ for any scalar r is also a solution of $Ax = b$. This is the case since $A(x_1 + rx) = Ax_1 + rAx = b + ro = b$. Thus, if $Ax = b$ has more than one solution, then it must have infinitely many solutions.

Unit Two Matrix Algebra

SOME FACTS ABOUT NONSINGULAR MATRICES (LESSON 2.3)

Fact 2.1

If the $n \times n$ matrix A has an inverse, then the inverse is unique.

Proof

Suppose that the matrix A has two distinct inverses B and C. Then the definition of the inverse of a matrix and the associative property of multiplication imply that

$$B = BI_n = B(AC) = (BA)C = I_nC = C$$

where I_n is the identity matrix.

Fact 2.2

If A is $n \times n$ invertible matrix, then $(A^{-1})^{-1} = A$.

Proof

Let $B = A^{-1}$. From the definition of the inverse $BA = I_n$. Multiply by B^{-1} to get $B^{-1}(BA) = B^{-1}I_n = B^{-1}$. This implies that $A = B^{-1}$. Hence, $(A^{-1})^{-1} = A$.

Fact 2.3

If A and B are $n \times n$ invertible matrices, then $(AB)^{-1} = B^{-1}A^{-1}$.

Proof

By employing the associative property of multiplication and the fact that the inverse of a matrix is unique, we get

$$(AB)(B^{-1}A^{-1}) = ((AB)B^{-1})A^{-1} = (A(BB^{-1}))A^{-1} = A(I_n)A^{-1} = AA^{-1} = I_n$$

Therefore, $(AB)^{-1} = B^{-1}A^{-1}$.

Fact 2.4

An $n \times n$ matrix A is invertible if and only if it is row equivalent to the identity matrix I_n.

Proof

On augmenting matrix A with identity $[A : I_n]$ and applying the Gauss-Jordan elimination algorithm, we observe that matrix A is invertible if and only if the reduced row echelon form of $[A : I_n]$ is transformed to $[I_n : B]$; that is, the matrix A is premultiplied by a sequence of elementary matrices say, $E_1, E_2, E_3, \ldots, E_r$ to yield the identity matrix. In this case $E_r, E_{r-1}, E_{r-2}, \ldots, E_2E_1A = I_n$. Hence, A is invertible if and only if A reduces to the identity matrix I_n.

NONSINGULAR MATRICES AND SYSTEMS OF LINEAR EQUATIONS (LESSON 2.3)

Fact 2.5

If $Ax = b$ is a system of n linear equations with n unknowns and if A is a nonsingular matrix, then the system has the unique solution $x = A^{-1}b$.

Proof

We will first show that $x = A^{-1}b$ is a solution to the system. This is the case since $Ax = AA^{-1}b = (AA^{-1})b = I_nb = b$. Next we will show that the solution is unique. Assume that x_1 and x_2 are two distinct solutions of the system. Then $Ax_1 = b$ and $Ax_2 = b$. Hence, $A(x_1 - x_2) = 0$. Since A is nonsingular, it follows, on multiplying by A^{-1}, that $x_1 - x_2 = 0$. This contradicts the assumption that these are two distinct solutions. Thus the solution is unique.

Fact 2.6

From Fact 2.5 we can deduce that if, for the homogeneous system $Ax = 0$ of n linear equations with n unknowns, the $n \times n$ coefficient matrix A is nonsingular, then the system has a unique solution (the zero solution).

SOME FACTS ABOUT DETERMINANTS (LESSON 2.5)

Fact 2.7

Let A and B be any $n \times n$ matrices and k be any nonzero scalar. The following properties hold:

a. $\det(AB) = \det(A)\det(B)$.

b. $\det(A^T) = \det(A)$, where A^T is the transpose of A.

c. If A is a nonsingular matrix, then $\det(A^{-1}) = \frac{1}{\det(A)}$

d. If B is the matrix obtained from A by multiplying one row (column) of A by k, then $\det(B) = k * \det(A)$.

e. If B is the matrix obtained from A by multiplying A by k, then $\det(B) = k^n \det(A)$.

f. If B is the matrix obtained from A by interchanging two rows (columns) of A, then $\det(B) = -\det(A)$.

g. If B is the matrix obtained from A by adding a multiple of one row (column) to another row (column), then $\det(B) = \det(A)$.

h. If any two rows (columns) of a matrix A are identical, then $\det(A) = 0$.

i. If any row (column) of a matrix A is a zero row, then $\det(A) = 0$.

j. If any two rows (columns) of a matrix A are proportional, then $\det(A) = 0$.

Proof of (c)

From (a) and the fact that $A^{-1}A = I_n$, it follows that

$$\det(A^{-1}) = \frac{1}{\det(A)}$$

Proof of (d)

Recall that every elementary row operation applied to matrix A can be performed by premultiplying the matrix A by the corresponding elementary matrix. Suppose that the ith row of A is multiplied by a constant k. This is equivalent to writing $B = EA$, where E is the elementary matrix obtained by multiplying the ith row of the identity matrix I_n by the nonzero scalar k. Since $\det(E) = k$, it follows that

$$\det(B) = \det(EA) = \det(E)\det(A) = k * \det(A).$$

Proof of (e)

This is an immediate consequence of (d).

Proof of (f)

If the ith row of A is interchanged with its jth row to obtain the matrix B, then this is equivalent to writing $B = EA$, where E is the elementary matrix obtained by interchanging the ith row and the jth row of the identity matrix I_n. Since $\det(E) = -1$ ($E^2 = I$), then

$$\det(B) = \det(EA) = \det(E)\det(A) = -\det(A)$$

Proof of (g)

If a matrix B is obtained from matrix A by replacing the jth row of A by the sum of a multiple of the ith row of A and the jth row of A, then this is equivalent to writing $B = EA$, where E is the elementary matrix obtained by replacing the jth row of I_n by the sum of a multiple of the ith row of I_n and the jth row of I_n. Since $\det(E) = 1$ (E is a triangular matrix), it follows that

$$\det(B) = \det(EA) = \det(E)\det(A) = \det(A).$$

DETERMINANTS AND NONSINGULAR MATRICES (LESSONS 2.5 AND 2.6)

Fact 2.8

An $n \times n$ matrix A is nonsingular if and only if $\det(A) \neq 0$.

Proof

Assume that A is a nonsingular matrix. Then $A^{-1}A = I_n$ implies that $\det(A^{-1}A) = \det(A^{-1})\det(A) = 1$. Hence $\det(A) \neq 0$. Conversely, assume that $\det(A)$ is not equal to zero. The sequence of elementary matrices E_i that results from transforming A into its reduced echelon form is such that $\det(E_i) \neq 0$ for each i since $\det(A) \neq 0$. Thus the reduced echelon form of A has a nonzero determinant. Since the only $n \times n$ reduced echelon form with nonzero determinant is the identity matrix, it follows that matrix A can be reduced by a sequence of elementary row operations to the identity. Thus A is a nonsingular matrix.

Fact 2.9

Let A be an $n \times n$ nonsingular matrix. Then the A^{-1} is given by

$$A^{-1} = \frac{1}{\det(A)} Adj(A)$$

Proof

Using the cofactor expansion across the ith row, we obtain

$$\det(A) = a_{i1}C_{i1} + a_{i2}C_{i2} + a_{i3}C_{i3} + \ldots + a_{in}C_{in}$$

where C_{ij} is the cofactor of the entry A_{ij}. Let B be an $n \times n$ matrix obtained from A by replacing the ith and jth rows by the ith row of A. Then using the cofactor expansion across the jth row of B yields

$$\det(B) = a_{i1}C_{j1} + a_{i2}C_{j2} + a_{i3}C_{j3} + \ldots + a_{in}C_{jn}$$

Since $\det(B) = 0$, it follows that

$$a_{i1}C_{j1} + a_{i2}C_{j2} + a_{i3}C_{j3} + \ldots + a_{in}C_{jn} = 0$$

Consider the matrix $A \times Adj(A) = (d_{ij})$. Then

$$d_{ij} = a_{i1}C_{j1} + a_{i2}C_{j2} + a_{i3}C_{j3} + \ldots + a_{in}C_{jn}$$

We note that if $i = j$, then $d_{ij} = \det(A)$ otherwise 0. Therefore,

$$A \times Adj(A) = \det(A)I_n$$

Since A is nonsingular, multiply both sides by A^{-1} to obtain the result

$$A^{-1} = \frac{1}{\det(A)} Adj(A).$$

DETERMINANTS AND SYSTEMS OF LINEAR EQUATIONS (LESSON 2.5)

Fact 2.10

Assume that $Ax = b$ is a system of n linear equations with n unknowns. If $\det(A)$ is not zero, then the system has a unique solution; otherwise, the system has infinitely many solutions or no solution.

Proof

If $\det(A) \neq 0$, then A is nonsingular. Fact 2.5 implies that the solution is unique and is given by $x = A^{-1}b$. On the other hand, if $\det(A) = 0$, then the reduced echelon form of matrix A cannot be identity matrix I_n. In this case, the system cannot have a unique solution. Thus, the system may have no solutions or infinitely many solutions.

Fact 2.11 (Cramer's Rule)

Assume $Ax = b$ is a system of n linear equations in n unknowns. If $\det(A) \neq 0$, the unique solution of the system is given by $x_i = \frac{\det(A_i)}{\det(A)}$, $i = 1, 2, 3, \ldots, n$ where A_i is the matrix obtained by replacing the ith column of A by b.

Proof

Since matrix A is nonsingular, it follows that the solution is given by $x = A^{-1}b$. Thus,

$$x = \frac{1}{\det(A)} Adj(A) * b.$$

Hence, the solution is

$$x_i = \frac{1}{\det(A)}(b_1 C_{1i} + b_2 C_{2i} + b_3 C_{3i} + \ldots + b_n C_{ni}), i = 1, 2, \ldots, n,$$

where C_{ij} is the ijth entry of $\text{Adj}(A)$. But $b_1 C_{1i} + b_2 C_{2i} + b_3 C_{3i} + \ldots + b_n C_{ni} = \det(A_i)$ and thus $x_i = \frac{\det(A_i)}{\det(A)}$.

PROOF OF FACTS — Unit Three Linear Spaces

SUBSPACES (LESSON 3.3)

Fact 3.1

If S is a subspace of a given linear space V, then the zero vector **0** is in S.

Proof:

Let v be a vector in S and 0 be a scalar. Since S is closed under scalar multiplication, it follows that $\mathbf{0} = 0 * v$ in S.

LINEAR COMBINATIONS AND SPANNING (LESSON 3.2)

Fact 3.2

Let $S = \{v_1, v_2, v_3, \ldots, v_n\}$ be a subset of a linear space V. Then the set of all linear combinations of S,

$$L(S) = \{a_1 v_1 + a_2 v_2 + a_3 v_3 + \ldots + a_n v_n \mid a_1, a_2, a_3, \ldots, a_n \text{ in } R\}$$

is a subspace of V.

Proof

We need to show that S is closed under both addition and scalar multiplication. Let

$$u_1 = a_1 v_1 + a_2 v_2 + a_3 v_3 + \ldots + a_n v_n$$

and

$$u_2 = b_1 v_1 + b_2 v_2 + b_3 v_3 + \ldots + b_n v_n$$

be any two elements of L(S). Then

$$u_1 + u_2 = (a + b_1)v_1 + (a_2 + b_2)v_2 + (a_3 + b_3)v_3 + \ldots + (a_n + b_n)v_n$$

Since each $a_i + b_i \in R$ for $i = 1, 2, 3, \ldots, n$, it follows that $u_1 + u_2$ is a linear combination of $v_1, v_2, v_3, \ldots, v_n$ and hence it belongs to $L(S)$. Thus $L(S)$ is closed under addition. If c is any scalar, then

$$cu_1 = c(a_1 v_1 + a_2 v_2 + a_3 v_3 + \ldots + a_n v_n) = ca_1 v_1 + ca_2 v_2 + ca_3 v_3 + \ldots + ca_n v_n$$

Since each $c\,a_i \in R$ for $i = 1, 2, 3, \ldots, n$, it follows that cu_1 is a linear combination of $v_1, v_2, v_3, \ldots, v_n$ and hence it belongs to $L(S)$. That is, $L(S)$ is closed under scalar multiplication. Since $L(S)$ is closed under addition and scalar multiplication, it follows that $L(S)$ is a subspace of V spanned by the set $S = \{v_1, v_2, v_3, \ldots, v_n\}$.

SYSTEMS OF LINEAR EQUATIONS AND LINEAR COMBINATIONS (LESSON 3.2 AND LESSON 3.4)

Fact 3.3

A vector w in R^m is a linear combination of the set of vectors $\{v_1, v_2, v_3, \ldots, v_n\}$ in R^m if and only if the associated nonhomogeneous system of $m \times n$ linear equations $Ac = b$ has nontrivial solutions.

Proof

The vector $w = [b_1, b_2, b_3, \ldots, b_m]$ is a linear combination of set of vectors $\{v_1, v_2, v_3, \ldots, v_n\} \in R^m$ if and only if there exist scalars $c_1, c_2, c_3, \ldots, c_n$ such that

$$w = c_1 v_1 + c_2 v_2 + c_3 v_3 + \ldots + c_n v_n$$

Since the vector $v_i = [a_{i1}, a_{i2}, a_{i3}, \ldots, a_{im}]$ for $i = 1, 2, \ldots, n$, we have the nonhomogeneous system

$$c_1 a_{11} + c_2 a_{21} + c_3 a_{31} + \ldots + c_n a_{n1} = b_1$$
$$c_1 a_{12} + c_2 a_{22} + c_3 a_{32} + \ldots + c_n a_{n2} = b_2$$
$$\vdots$$
$$c_1 a_{1m} + c_2 a_{2m} + c_3 a_{3m} + \ldots + c_n a_{nm} = b_m$$

This is equivalent to the matrix equation $Ac = b$ where A is the $m \times n$ coefficient matrix $[a_{ij}]$ and c is the $n \times 1$ matrix $[c_1, c_2, c_3, \ldots, c_n]$. A vector w is a linear combination of the vectors $v_1, v_2, v_3, \ldots, v_n$ if and only if the nonhomogeneous system has nontrivial solutions.

Fact 3.4

If at least one vector of set $S = \{v_1, v_2, v_3, \ldots, v_n\}$ of a linear space V is a linear combination of the remaining vectors of S, then S is a dependent set.

Proof

Let v_i, for some i, be a linear combination of $\{v_1, v_2 \ldots, v_{i-1}, v_{i+1} \ldots, v_n\}$. Then there exist scalars $c_1, c_2 \ldots, c_{i-1}, c_{i+1} \ldots, c_n$ such that

$$v_i = c_1 v_1 + c_2 v_2 + c_{i-1} v_{i-1} + c_{i+1} v_{i+1} \ldots + c_n v_n$$

Thus there exist scalars not all zero (in particular the coefficient of v_i) such that

$$c_1 v_1 + c_2 v_2 + c_{i-1} v_{i-1} + (-1) v_i + c_{i+1} v_{i+1} \ldots + c_n v_n = 0$$

Therefore, by definition, set S is linearly dependent.

Fact 3.5

If set $S = \{0, v_1, v_2, v_3, \ldots, v_n\}$ of a linear space V includes the zero vector, then S is a dependent set.

Proof

Let $c, c_1, c_2, \ldots, c_n$ be scalars and consider the zero combination of the elements of S:

$$c.0 + c_1 v_1 + c_2 v_2 \ldots + c_n v_n = 0$$

Clearly, this combination holds if all the c_i's are equal to zero and c is any nonzero scalar. Thus there exists a nonzero scalar (c in this case) for which

$$c.0 + c_1 v_1 + c_2 v_2 \ldots + c_n v_n = 0$$

Therefore S is a dependent set.

Fact 3.6

If S_1 is a nonempty subset of a linearly independent set $S = \{v_1, v_2, v_3, \ldots, v_n\}$ of a linear space V, then S_1 must be an independent set.

Proof

Assume that set $S_1 = \{v_1, v_2, v_3, \ldots, v_k\}$ in S is a dependent set. Then there exist scalars $c_1, c_2, \ldots, c_k$ not all zero such that $c_1 v_1 + c_2 v_2 + \ldots + c_k v_k = 0$. If we choose scalars $c_{k+1} = 0, c_{k+2} = 0 \ldots, c_n = 0$, we have

$$c_1 v_1 + c_2 v_2 + \ldots + c_k v_k + c_{k+1} v_{k+1} + c_{k+2} v_{k+2} + \ldots + c_n v_n = 0$$

with not all scalars being zero. This will imply that set S is linearly dependent. This contradiction proves that subset S_1 must be independent.

HOMOGENEOUS SYSTEMS OF LINEAR EQUATIONS AND DEPENDENT/INDEPENDENT SET (LESSON 3.4)

Fact 3.7

A set of vectors $\{v_1, v_2, v_3, \ldots, v_n\}$ in R^m is linearly dependent if and only if the associated homogeneous system of $m \times n$ linear equations $Ax = 0$ has nontrivial solutions.

Proof

The set $\{v_1, v_2, v_3, \ldots, v_n\}$ in R^m is a dependent set if and only if there exist scalars $c_1, c_2, \ldots, c_n$ not all zero such that $c_1 v_1 + c_2 v_2 + \ldots + c_n v_n = 0$. Since the vector v_i for $i = 1, 2, \ldots, n$ is $v_i = [a_{i1}, a_{i2}, a_{i3}, \ldots, a_{im}]$, we have the homogeneous system

$$c_1 a_{11} + c_2 a_{21} + c_3 a_{31} + \ldots + c_n a_{n1} = 0$$
$$c_1 a_{12} + c_2 a_{22} + c_3 a_{32} + \ldots + c_n a_{n2} = 0$$
$$\vdots$$
$$c_1 a_{1m} + c_2 a_{2m} + c_3 a_{3m} + \ldots + c_n a_{nm} = 0$$

This is equivalent to the matrix equation $Ac = 0$ where A is the $m \times n$ coefficient matrix $[a_{ij}]$ and c is the $n \times 1$ matrix $[c_1, c_2, \ldots, c_n]$. Now the scalars $c_1, c_2, \ldots, c_n$ are not all equal to zero if and only if the homogeneous system has nontrivial solutions.

Fact 3.8

A set of vectors $\{v_1, v_2, v_3, \ldots, v_n\}$ in R^n is linearly independent if and only if the associated homogeneous system of $n \times n$ linear equations $Ax = 0$ has a unique solution.

BASIS AND DIMENSION (LESSON 3.5)

Fact 3.9

Let $S = \{v_1, v_2, v_3, \ldots, v_n\}$ be a basis for a linear space V. Every set $S_1 = \{u_1, u_2, u_3, \ldots, u_m\}$ containing m vectors with $m > n$ must be dependent and hence cannot be a basis for V.

Proof

We want to show that there exist scalars not all zero such that

$$k_1 u_1 + k_2 u_2 + \ldots + k_m u_m = 0$$

Since S is a basis for the linear space V, it follows that each vector u_i, $i = 1, 2, \ldots, m$, can be expressed as a linear combination of the basic vectors $v_1, v_2, v_3, \ldots, v_n$. That is,

$$u_1 = c_{11} v_1 + c_{12} v_2 + \ldots + c_{1n} v_n$$
$$u_2 = c_{21} v_1 + c_{22} v_2 + \ldots + c_{2n} v_n$$
$$\vdots$$
$$u_m = c_{m1} v_1 + c_{m2} v_2 + \ldots + c_{mn} v_n.$$

On substituting these into $k_1 u_1 + k_2 u_2 + \ldots + k_m u_m = 0$, we get a homogeneous system of n equations with m unknowns $k_1, k_2, \ldots, k_m$. Since $m > n$, it follows that the homogeneous system has nontrivial solutions. Therefore, there exist scalars $k_1, k_2, \ldots, k_m$ not all zero such that

$$k_1 u_1 + k_2 u_2 + \ldots + k_m u_m = 0$$

and $S_1 = \{u_1, u_2, u_3, \ldots, u_m\}$ is dependent.

Fact 3.10

Let V be a linear space whose dimension is n. Then any set $S = \{v_1, v_2, v_3, \ldots, v_n\}$ of n independent vectors forms a basis for V.

Proof

It suffices to show that $L(S) = V$. Since $L(S)$ is a subset of V, we need only show that V is a subset of $L(S)$. Let v in V. Since the dimension of V is n, Fact 3.9 implies that the set $\{v, v_1, v_2, v_3, \ldots, v_n\}$ is a dependent set. Hence there exist scalars $c, c_1, c_2, c_3, \ldots, c_n$ such that

$$cv + c_1 v_1 + c_2 v_2 + c_3 v_3 + \ldots + c_n v_n = 0 \ (c \neq 0)$$

and this implies that

$$v = (-c_1/c)v_1 + (-c_2/c)v_2 + \ldots + (-c_n/c)v_n$$

Therefore, $v \in L(S)$ and $V = L(S)$.

Fact 3.11

Let V be a linear space whose dimension is n. Then any set $S = \{v_1, v_2, v_3, \ldots, v_n\}$ of n vectors that spans the space V forms a basis for V.

Proof

It is enough to show that set $S = \{v_1, v_2, v_3, \ldots, v_n\}$ is linearly independent. If S were a dependent set, then one of the v_i's is a linear combination of the remaining vectors. This implies that a set of $(n-1)$ vectors spans a space of dimension n. Since this cannot be the case, it follows that set S is linearly independent.

Fact 3.12

A basis for a linear space V is not unique. All bases must have the same number of elements.

Proof

Let $S = \{v_1, v_2, v_3, \ldots, v_n\}$ and $U = \{u_1, u_2, u_3, \ldots, u_m\}$ be two bases for the linear space V. From Fact 3.9, it follows that (a) if S is a basis and U is an independent set of V, then $m \leq n$; (b) on the other hand, if U is a basis and S is an independent set of V, then $n \leq m$. Hence $n = m$. This shows that all bases must have the same number of elements.

Fact 3.13

Let $S = \{v_1, v_2, v_3, \ldots, v_n\}$ be a basis for a linear space V. Then every vector w in V can be expressed uniquely as a linear combination of the elements of S.

Proof

Assume that vector w can be expressed in two different ways as a linear combination of the vectors $v_1, v_2, v_3, \ldots, v_n$. Thus there exist scalars $c_1, c_2, c_3, \ldots, c_n$ and $d_1, d_2, d_3, \ldots, d_n$ such that

$$w = c_1 v_1 + c_2 v_2 + \ldots + c_n v_n$$

and

$$w = d_1 v_1 + d_2 v_2 + \ldots + d_n v_n$$

Hence,

$$c_1 v_1 + c_2 v_2 + \ldots + c_n v_n = d_1 v_1 + d_2 v_2 + \ldots + d_n v_n$$

This implies that

$$(c_1 - d_1)v_1 + (c_2 - d_2)v_2 + (c_3 - d_3)v_3 + \ldots + (c_n - d_n)v_n = 0$$

Since set $S = \{v_1, v_2, v_3, \ldots, v_n\}$ is a basis, the vectors $v_1, v_2, v_3, \ldots, v_n$ are linearly independent. It follows that $c_1 = d_1, c_2 = d_2, c_3 = d_3, \ldots, c_n = d_n$. Therefore, the vector w can be expressed uniquely as a linear combination of its basis elements.

ROW AND COLUMN SPACE OF A MATRIX (LESSON 3.6)

Fact 3.14

The row space and the column space of an $m \times n$ matrix A span subspaces of R^n and R^m respectively.

Since elementary row operations do not alter the row space of a matrix, we have

Fact 3.15

The nonzero rows of an $m \times n$ matrix A in echelon form constitute a basis for the row space of A.

Fact 3.16

The row space and the column space of an $m \times n$ matrix A have the same dimension. This common number defines the rank of matrix A.

Proof

Let $r_i = (a_{i1}, a_{i2}, \ldots, a_{in})$ be the ith row of matrix A, $i = 1, 2, \ldots, m$. Let us assume that the dimension of the row space is p and that the vectors $v_i = (b_{i1}, b_{i2}, \ldots, b_{in})$, $i = 1, 2, \ldots, p$, be a basis for the row space. Then each row of A can be expressed as a linear combination of the v's so that for each i we have $r_i = c_{i1}v_1 + c_{i2}v_2 + \ldots + c_{ip}v_p$. This implies that, for $j = 1, 2, \ldots, n$, we have

$$a_{1j} = c_{11}b_{1j} + c_{12}b_{2j} + \ldots + c_{1p}b_{pj},$$
$$a_{2j} = c_{21}b_{1j} + c_{22}b_{2j} + \ldots + c_{2p}b_{pj}$$
$$\vdots$$
$$a_{mj} = c_{m1}b_{1j} + c_{m2}b_{2j} + \ldots + c_{mp}b_{pj}.$$

This computation in turns implies that each column of matrix A is a linear combination of p columns of the form $c_i = (c_{1i}, c_{2i}, \ldots, c_{mi})$, $i = 1, 2, \ldots, p$. Thus dim(column space) $\leq$ dim(row space). Similarly, one can show that dim(row space) $\geq$ dim(column space).

Fact 3.17

An $n \times n$ matrix A is nonsingular if and only if rank(A) $= n$.

Proof

Matrix A is nonsingular if and only if the matrix A is row equivalent to the identity if and only if A has n independent rows if and only if rank(A) $= n$.

Fact 3.18

Let A be an $m \times n$ matrix and $Ax = b$ be a linear system of m equations in n unknowns. Then the system has a solution if and only if the rank of the augmented matrix, rank[A : b], is equal to rank[A] of the coefficient matrix.

Proof

The system $Ax = b$ has a solution if and only if the column vector b belongs to the column space of A. Thus, if the vector b is added to the columns of the matrix A, the rank of the augmented matrix will not change. If this is the case, then the rank of the coefficient matrix and the rank of the augmented matrix are equal.

Unit Four   Inner Product Spaces

PROPERTIES OF THE SCALAR (DOT/INNER) PRODUCT (LESSON 4.1)

Fact 4.1

Let u, v, and w be three vectors in R^n and k be a real scalar. Then the following properties hold:

1. $u.v = v.u$
2. $u.u \geq 0$ and $u.u = 0$ if and only if $u = 0$.
3. $(ku).v = u.(kv) = k(u.v)$
4. $u.(v + w) = u.v + u.w$

Proof

Let $u = [u_1, u_2, \ldots, u_n]$, $v = [v_1, v_2, \ldots, v_n]$, and $w = [w_1, w_2, \ldots, w_n]$ be three vectors in R^n. Then, the standard inner product in R^n implies that

1. $u.v = u_1 v_1 + u_2 v_2 + \ldots + u_n v_n = v_1 u_1 + v_2 u_2 + \ldots + v_n u_n = v.u$
2. $u.u = u_1 u_1 + u_2 u_2 + \ldots + u_n u_n = (u_1)^2 + (u_2)^2 + \ldots + (u_n)^2 \neq 0$ unless $u = [0, 0, \ldots, 0]$ is the zero vector.
3.
$$(ku).v = [ku_1, ku_2, \ldots, ku_n].[v_1, v_2, \ldots, v_n]$$
$$= (ku_1)v_1 + (ku_2)v_2 + \ldots + (ku_n)v_n$$
$$= u_1(kv_1) + u_2(kv_2) + \ldots + u_n(kv_n)$$
$$= u.(kv) = k(u_1 v_1 + u_2 v_2 + \ldots + u_n v_n) = k(u.v)$$

4.
$$u.(v + w) = [u_1, u_2, \ldots, u_n].[v_1 + w_1, v_2 + w_2, \ldots, v_n + w_n]$$
$$= u_1(v_1 + w_1) + u_2(v_2 + w_2) + \ldots + u_n(v_n + w_n)$$

$$= u_1v_1 + u_1w_1 + u_2v_2 + u_2w_2 + \ldots + u_nv_n + u_nw_n$$

$$= u_1v_1 + u_2v_2 + \ldots + u_nv_n + u_1w_1 + u_2w_2 + \ldots + u_nw_n$$

$$= u.v + u.w.$$

Fact 4.2 (Cauchy-Schwartz inequality)

Let u and v be two vectors in R^n. Then the following inequality holds: $|u.v| \le \|u\|\|v\|$ where $\|v\|$ denotes the norm (length, magnitude) of the vector v. The equality holds provided that the two vectors are linearly dependent.

Proof

If one of the vectors is a multiple of the other vector, equality holds and there is nothing to prove. Assume that the vectors are nonzero vectors. Let a be a real number and construct the vector $w = au + v$. Since the product $w.w \ge 0$, this implies that

$$a^2(u.u) + 2a(u.v) + (v.v) \ge 0.$$

Viewing this as a quadratic inequality in a and since the coefficient of the quadratic term a^2 is always positive, the discriminant $4(u.v) - 4(u.u)(v.v) \le 0$. This observation implies that $|u.v| \le \|u\|\|v\|$.

Fact 4.3 (Triangle inequality)

Let u and v be two vectors in R^n. Then the following inequality holds: $\|u+v\| \le \|u\|+\|v\|$ and the equality holds provided the two vectors are linearly dependent.

Proof

From the properties of the inner product and the Cauchy-Schwartz inequality,

$$\|u+v\|^2 = (u+v).(u+v) = (u.u)+2(u.v)+(v.v) \le \|u\|^2+2\|u\|\|v\|+\|v\|^2 = (\|u\|+\|v\|)^2.$$

This computation yields the desired inequality.

ORTHOGONAL SETS AND GRAM-SCHMIDT PROCESS (LESSON 4.2 AND 4.3)

Fact 4.4

The orthogonal projection of a vector v upon a vector u, denoted by $\text{proj}_u v$, is given by $\text{proj}_u^v = (\frac{v.u}{u.u})u$ and $v - \text{proj}_u v$ is orthogonal to u.

Proof

We will only verify that u is orthogonal to v-$\text{proj}_u v$. Compute the inner product of the two vectors:

$$u.(v\text{-proj}_u v) = u.v - \left(\frac{v.u}{u.u}\right) u.u = u.v - v.u = 0$$

Fact 4.5

If the set $S = \{v_1, v_2, \ldots, v_n\}$ is an orthonormal basis for a linear space V, then every vector $v \in V$ can be expressed

$$v = (v.v_1)v_1 + (v.v_2)v_2 + \ldots + (v.v_n)v_n$$

Proof

Since the set $\{v_1, v_2, \ldots, v_n\}$ forms a basis, it follows that every vector $v \in V$ van be expressed $v = c_1 v_1 + c_2 v_2 + c_3 v_3 + \ldots + c_n v_n$. Since the set $\{v_1, v_2, \ldots, v_n\}$ is orthonormal, it follows that $v_i.v = c_i(v_i.v_i) = c_i$. This proves that vector v can be expressed

$$v = (v.v_1)v_1 + (v.v_2)v_2 + \ldots + (v.v_n)v_n.$$

Fact 4.6

If $S = \{v_1, v_2, \ldots, v_n\}$ is a nonempty subset consisting of nonzero mutually orthogonal vectors, then set S is a linearly independent set.

Proof

Consider the linear combination $c_1 v_1 + c_2 v_2 + c_3 v_3 + \ldots + c_n v_n = 0$ of the elements of the S. For each $i = 1, 2, \ldots n$, the product

$$v_i.(c_1 v_1 + c_2 v_2 + c_3 v_3 + \ldots + c_n v_n) = 0$$

Since the vectors are mutually orthogonal, it follows that $c_i(v_i.v_i) = 0$. Thus, for each i, $c_i = 0$, proving that the set of vectors $\{v_1, v_2, \ldots, v_n\}$ is linearly independent.

Fact 4.7

Let a set $S = \{v_1, v_2, \ldots, v_n\}$ be a basis for a linear space V. Then the process of constructing an orthonormal basis $O = \{u_1, u_2, \ldots, u_n\}$ from the basis S is $u_1 = \frac{v_1}{\|v_1\|}$, $w_2 = v_2 - \text{proj}_{u_1} v_2$ and $u_2 = \frac{w_2}{\|w_2\|} \ldots$, and in general

$$w_n = v_n - \text{proj}_{u_{n-1}} v_n - \text{proj}_{u_{n-2}} v_n - \ldots - \text{proj}_{u_1} v_n; \; u_n = \frac{w_n}{\|w_n\|}$$

Proof

Clearly u_1 is a unit vector. The orthogonal projection of u_2 on u_1 implies that the vector w_2 is orthogonal to u_1. Thus, $\{u_1, u_2\}$ are the first two elements of the orthonormal basis. Now we construct a third vector w_3, which is orthogonal to both u_1 and u_2. The vector $w_3 = v_3 - \text{proj}_{u_2} v_3 - \text{proj}_{u_1} v_3$ is orthogonal to both u_1 and u_2. Thus, the set $\{u_1, u_2, u_3\}$ consists of the first three elements of the orthonormal basis. We repeat this process until we exhaust all the elements of set S. The constructed set is the required orthonormal basis.

ORTHOGONAL MATRICES AND THE QR-DECOMPOSITION (LESSON 4.4)

Fact 4.8

The following properties hold for any orthogonal matrix A:

- The rows of A form an orthonormal set.
- The inverse of A exists and $A^{-1} = A^T$.
- $\det(A) = +1$ or -1.

Unit Five   Linear Transformations

ALGEBRA OF LINEAR TRANSFORMATIONS (LESSON 5.1)

Let T_1 and T_2 be two linear transformations defined from a linear space U into a linear space V. The sum and the scalar multiplication of these transformations can be defined as

$$(T_1 + T_2)(u) = T_1(u) + T_2(u), \ (cT_1)(u) = cT_1(u)$$

for any u in U and a real scalar c.

Fact 5.1

Let $T_1 : U \to V$, $T_2 : U \to V$, and $T_3 : U \to V$ be three linear transformations defined from a linear space U into a linear space V. The following properties hold:

- $T_1 + T_2 = T_2 + T_1$
- $(T_1 + T_2) + T_3 = T_1 + (T_2 + T_3)$
- $T_1 + 0 = 0 + T_1 = T_1$, where 0 is the transformation that maps every vector in U to zero in V
- $T_1 + (-T_1) = 0$,
- $c(T_1 + T_2) = cT_1 + cT_2$
- $(c_1 + c_2)T_1 = c_1T_1 + c_2T_1$
- $1.T_1 = T_1$

Fact 5.2

Let $T : U \to V$ be a linear transformation from a linear space U into a linear space V. Then $T(0) = 0$. That is, if T is a linear transformation, then it maps the zero element of U to the zero element of V. The converse of this statement is not true.

Proof

Since T is a linear transformation, $T(0) = T(0 + 0) = T(0) + T(0)$. Thus $T(0) = 0$.

Let $T_1 : W \to V$ and $T_2 : V \to W$ be two linear transformations from the linear spaces W to V and from V into W respectively. The composition of the transformations, $T_1 \circ T_2$, from V into V is defined

$$(T_1 \circ T_2)(v) = T_1(T_2(v))$$

Fact 5.3

The composition of linear transformations is a linear transformation.

Proof

If $T_1 : W \to V$ and $T_2 : V \to W$ are two linear transformations from the linear space W to V and from V into W, respectively, then for any vectors v_1 and v_2 in V and any real scalars a and b,

$$(T_1 \circ T_2)(av_1 + bv_2) = T_1(T_2(av_1 + bv_2))$$

Since T_2 is a linear transformation, it follows that

$$(T_1 \circ T_2)(av_1 + bv_2) = T_1(aT_2(v_1) + bT_2(v_2)))$$

Since T_1 is a linear transformation, it follows that

$$(T_1 \circ T_2)(av_1 + bv_2) = aT_1(T_2(v_1)) + bT_1(T_2(v_2))$$
$$= a(T_1 \circ T_2)(v_1) + b(T_1 \circ T_2)(v_2)$$

The composition is a linear transformation.

KERNEL AND RANGE OF A LINEAR TRANSFORMATION (LESSON 5.1 AND LESSON 5.2)

Fact 5.4

Let $T : U \to V$ be a linear transformation. Let $B = \{u_1, u_2, \ldots, u_n\}$ be a basis for the linear space U.

1. The action of T on U is determined by its action on the basis set B.
2. $\{T(u_1), T(u_2), \ldots, T(u_n)\}$ is a basis for range(T) if and only if ker(T) = {0}.

Proof

1. Since $T(u_i)$ is defined for each $i = 1, 2, \ldots, n$ and every vector is a linear combination of elements of the basis set B, $u = a_1 u_1 + a_2 u_2 + \ldots + a_n u_n$, it follows that T(u) is completely determined.

2. Clearly set $\{T(u_1), T(u_2), \ldots, T(u_n)\}$ spans range(T). We need to verify that this set is linearly independent. Let

$$a_1 T(u_1) + a_2 T(u_2) + \ldots + a_n T(u_n) = 0.$$

Since T is a linear transformation, it follows that $T(a_1 u_1 + a_2 u_2 + \ldots + a_n u_n) = 0$, which implies that $a_1 u_1 + a_2 u_2 + \ldots + a_n u_n$ is in ker(T). Now ker(T) = $\{0\}$ implies that $a_1 u_1 + a_2 u_2 + \ldots + a_n u_n = 0$. Since the vector set $\{u_1, u_2, \ldots, u_n\}$ is a basis, it follows that $a_1 = a_2 = \ldots = a_n = 0$. Thus, the set $\{T(u_1), T(u_2), \ldots, T(u_n)\}$ is linearly independent and forms a basis for range(T).

If $u \in$ ker(T) is a nonzero vector, then

$$0 = T(u) = T(a u_1 + a_2 u_2 + \ldots + a_n u_n)$$

$$= a_1 T(u_1) + a_2 T(u_2) + \ldots + a_n T(u_n)$$

Since the set $\{T(u_1), T(u_2), \ldots, T(u_n)\}$ is a basis for range(T), it follows that $a_1 = a_2 = \ldots = a_n = 0$. Therefore, ker(T) cannot contain any nonzero element and ker(T) = $\{0\}$.

Fact 5.5

Let $T : U \to V$ be a linear transformation. Then the following statements are true:

a. ker(T) is a subspace of U.
b. range(T) is a subspace of V

Proof

a. We need to show that ker(T) is closed under addition and scalar multiplication. Observe that ker(T) is a nonempty subset of U, since it contains 0 (why?). If u_1 and u_2 are any two elements of ker(T), then $T(u_1) = 0$ and $T(u_2) = 0$. Since T is a linear transformation, $T(u_1 + u_2) = T(u_1) + T(u_2) = 0$. Also, if c is any scalar, then $T(c u_1) = c T(u_1)) = 0$. Thus, $u_1 + u_2$ and $c u_1$ belong to ker(T). Therefore, ker(T) is a subspace of U.

b. We need to show that range(T) is closed under addition and scalar multiplication. Observe that range(T) is a nonempty subset of V since it contains 0 (why?). If v_1 and v_2 are any two elements of range(T), then there exist u_1 and u_2 in U such that $T(u_1) = v_1$ and $T(u_2) = v_2$. Since T is a linear transformation, $T(u_1 + u_2) = T(u_1) + T(u_2) = v_1 + v_2$. Also, if c is any scalar, then $T(c u_1) = c T(u_1)) = c v_1$. Thus, $v_1 + v_2$ and $c v_1$ belong to range(T). Therefore, range(T) is a subspace of V.

Fact 5.6 (Dimension Theorem)

Let $T : U \to V$ be a linear transformation. Then

$$\dim(\ker(T)) + \dim(\text{range}(T)) = \dim(U)$$

Proof

Suppose that $B = \{u_1, u_2, \ldots, u_n\}$ is a basis for U. If $ker(T) = \{0\}$, then Fact 5.4 implies that the set $\{T(u_1), T(u_2), \ldots, T(u_n)\}$ is a basis for range(T). In this case, the theorem holds with $dim(ker(T)) = 0$, $dim(range(T)) = dim(U) = n$.

Assume now that $dim(ker(T)) = m$. If $m = n$, then $ker(T) = U$ and in this case T is the zero transformation and hence $range(T) = \{0\}$. Consider now the case $1 \leq m < n$. Let $B_1 = \{u_1, u_2, \ldots, u_m\}$ be a basis for ker(T). We are done if we verify that the set $T_1 = \{T(u_{m+1}), T(u_{m+2}), \ldots, T(u_n)\}$ is a basis for range(T). First let us show that T_1 spans range(T). Let $v \in range(T)$. Then there is a vector $u \in U$ such that $T(u) = v$. Since B is a basis for the linear space U, it follows that

$$u = a_1 u_1 + a_2 u_2 + \ldots + a_m u_m + a_{m+1} u_{m+1} + \ldots + a_n u_n$$

Now

$$v = T(u) = T(a_1 u_1 + a_2 u_2 + \ldots + a_m u_m + a_{m+1} u_{m+1} + \ldots + a_n u_n)$$
$$= a_1 T(u_1) + a_2 T(u_2) + \ldots + a_m T(u_m) + a_{m+1} T(u_{m+1}) + \ldots + a_n T(u_n)$$
$$= a_{m+1} T(u_{m+1}) + \ldots a_n T(u_n).$$

Thus set T_1 spans range(T). Set T_1 is linearly independent. This is the case because

$$a_{m+1} T(u_{m+1}) + \ldots + a_n T(u_n) = 0$$

implies

$$T(a_{m+1} u_{m+1} + \ldots + a_n u_n) = 0$$

which in turn implies

$$a_{m+1} u_{m+1} + \ldots + a_n u_n \in ker(T)$$

Thus

$$a_{m+1} u_{m+1} + \ldots + a_n u_n = a_1 u_1 + a_2 u_2 + \ldots + a_{m-1} u_{m-1} + a_m u_m$$

implies

$$-a_1 u_1 - a_2 u_2 - \ldots - a_{m-1} u_{m-1} - a_m u_m + a_{m+1} u_{m+1} + \ldots + a_n u_n = 0$$

Since B is a basis for U, it follows that $a_1 = a_2 = \ldots = a_n = 0$. This proves that the set T_1 is a basis for range(T). Thus,

$$dim(ker(T)) + dim(range(T)) = dim(U).$$

Fact 5.7

A linear transformation $T : U \rightarrow V$ is one-to-one if and only if $ker(T) = \{0\}$.

Proof

For any vectors $u_1, u_2 \in U$, $T(u_1) = T(u_2) \implies T(u_1 - u_2) = 0 \implies u_1 - u_2 \in ker(T) \implies u_1 = u_2$.

Fact 5.8

Let T : U → U be a linear transformation. Then the following statements are equivalent:

- $\ker(T) = \{0\}$
- T is one-to-one
- inverse(T) exists

MATRIX REPRESENTATION OF A LINEAR TRANSFORMATION (LESSON 5.3)

If U is a linear space with a basis $E = \{e_1, e_2, \ldots, e_n\}$, then any vector $u \in U$ can be expressed uniquely as $u = u_1 e_1 + u_2 e_2 + \ldots + u_n e_n$. The coordinates of vector u relative to the basis E, denoted by u_E, are $[u_1, u_2, \ldots, u_n]$.

Fact 5.9

Let T : U → V be a linear transformation from the linear space U to the linear space V whose bases are $E = \{e_1, e_2, \ldots, e_n\}$ and $E_1 = \{s_1, s_2, \ldots, s_m\}$, respectively. Then there is an $m \times n$ matrix A that represents the transformation T relative to these bases.

Proof

Let u be any vector in U whose coordinates with respect to the basis E are $u_E = [u_1, u_2, \ldots, u_n]$; that is, $u = u_1 e_1 + u_2 e_2 + \ldots + u_n e_n$. Then

$$T(u) = T(u_1 e_1 + u_2 e_2 + \ldots + u_n e_n) = u_1 T(e_1) + u_2 T(e_2) + \ldots + u_n T(e_n)$$

Now each $T(e_i)$, for $i = 1, 2, \ldots n$, can be expressed uniquely in terms of the basis of V. Thus,

$$T(e_1) = a_{11} s_1 + a_{21} s_2 + \ldots + a_{m1} s_m$$
$$T(e_2) = a_{12} s_1 + a_{22} s_2 + \ldots + a_{m2} s_m$$
$$\vdots$$
$$T(e_n) = a_{1n} s_1 + a_{2n} s_2 + \ldots + a_{mn} s_m$$

Now form the $m \times n$ matrix whose columns are the coordinates relative to the basis E_1 of the vectors $T(e_i)$; that is, the ith column ($i = 1, 2, \ldots n$) is $[a_{1i}, a_{2i}, \ldots a_{ni}]$. This is the matrix that represents T.

CHANGE OF BASIS (LESSON 5.4)

Fact 5.10

Let U be a given linear space with two different bases $E = \{e_1, \ldots, e_n\}$ and $S = \{s_1, \ldots, s_n\}$. Let u be a vector in U whose coordinates relative to the bases E and S are u_E

and u_S respectively. Then these coordinates are related by the relation $u_S = Pu_E$ where P is the matrix whose columns consist of the coordinates $[e_i]_S$, $i = 1, 2, \ldots, n$ and is the transition matrix from the basis E to the basis S.

Proof

Each element of the basis E can be expressed uniquely in terms of the basis S. In particular,

$$e_1 = a_{11}s_1 + a_{21}s_2 + \ldots + a_{n1}s_n$$
$$e_2 = a_{12}s_1 + a_{22}s_2 + \ldots + a_{n2}s_n$$
$$\vdots$$
$$e_n = a_{1n}s_1 + a_{2n}s_2 + \ldots + a_{nn}s_n$$

Let u be a vector in U such that $u_E = [c_1, c_2, \ldots, c_n]$; that is,

$$u = c_1e_1 + c_2e_2 + \ldots + c_ne_n$$

Then

$$u = c_1(a_{11}s_1 + \ldots + a_{n1}s_n) + c_2(a_{12}s_1 + \ldots + a_{n2}s_n) + \ldots + c_n(a_{1n}s_1 + \ldots + a_{nn}s_n)$$
$$= (c_1a_{11} + \ldots + c_na_{1n})s_1 + (c_1a_{21} + \ldots + c_na_{2n})s_2 \ldots + (c_1a_{n1} + \ldots + c_na_{nn})s_n$$

Thus, the coordinates of the vector u relative to the basis S, u_S are equal to Pu_E, where P is the matrix whose ith column ($i = 1, 2, \ldots n$) is $[a_{i1}, a_{i2}, \ldots a_{in}]$.

Fact 5.11

The transition matrix P from the basis E to the basis S of Fact 5.10 is invertible and the inverse of P which is Q, represents the transition matrix from the basis S to the basis E.

Proof

Let u be any vector in U whose coordinates relative to the bases E and S are u_E and u_S respectively. Let P and Q represent the transition matrices from the basis E to the basis S and from the basis S to E, respectively. Then $u_S = Pu_E$ and $u_E = Qu_S$. Thus $u_S = PQu_S$ and $u_E = QPu_E$. Hence $PQ = QP = I_n$, I_n being the identity matrix. Therefore, P is invertible.

Fact 5.12

Let $T : U \rightarrow U$ be a linear transformation and U be a linear space with two different bases, $E = \{e_1, e_2, \ldots, e_n\}$ and $S = \{s_1, s_2, \ldots, s_n\}$. If A is the matrix representation of T relative to the basis E and B is the matrix representation of T relative to the basis S, and if P is the transition matrix from the basis S to the basis E, then $A = PBP^{-1}$.

Proof

Let u be any vector in U whose coordinates relative to the bases E and S are u_E and u_S respectively and are related by $u_E = Pu_S$. The coordinates of the vector $T(u)$ relative to the bases E and S are Au_E and Bu_S, respectively and are related via $Au_E = PBu_S$. This implies that $Au_E = PBP^{-1}u_E$. Since this holds for an arbitrary vector u with coordinates u_E, it follows that $A = PBP^{-1}$.

Unit Six   Eigenspaces

PROPERTIES OF EIGENVALUES AND EIGENVECTORS (LESSON 6.1)

Fact 6.1

Let A be an $n \times n$ matrix. Then the set S spanned by the eigenvectors corresponding to an eigenvalue λ of A is a subspace. This is called the eigenspace associated with the eigenvalue λ.

Proof

We need to verify that set S is closed under addition and scalar multiplication. Let v and u be two vectors in S. This implies that $Av = \lambda v$ and $Au = \lambda u$. Now,

$$A(v + u) = Av + Au = \lambda v + \lambda u = \lambda(v + u)$$

Hence $v + u$ is in S and S is closed under addition. Also for any real scalar c,

$$A(cv) = cA(v) = c(\lambda v) = \lambda(cv)$$

and this implies that cv is in S and hence S is closed under scalar multiplication. Therefore, set S is a subspace.

Fact 6.2

Let $T : U \to U$ be a linear transformation and U be a linear space with two different bases, $E = \{e_1, e_2, \ldots, e_n\}$ and $S = \{s_1, s_2, \ldots, s_n\}$. If A is the matrix representation of T relative to the basis E and B is the matrix representation of T relative to the basis S, then the eigenvalues of A and B are equal.

Proof

Let P be the transition matrix from basis S to basis E. It follows that matrices A and B are similar with $A = PBP^{-1}$. Thus,

$$\det(\lambda I_n - A) = \det(\lambda I_n - PBP^{-1}) = \det(P)\det(\lambda I_n - B)\det(P^{-1}) = \det(\lambda I_n - B)$$

This implies that the similar matrices A and B have the same eigenvalues.

Matrix Diagonalization (Lesson 6.2)

Fact 6.3

If an $n \times n$ matrix A has n independent eigenvectors, then A is diagonalizable.

Proof

Assume that A has the eigenvectors $v_1, v_2, \ldots, v_n$ corresponding to the eigenvalues $\lambda_1, \lambda_2, \ldots, \lambda_n$ (λ's may not be distinct). Construct the matrix P whose columns are these eigenvectors. Clearly, P is a nonsingular matrix. On computing the product AP, we get $AP = PD$, where D is a diagonal matrix with diagonal elements $\lambda_1, \lambda_2, \ldots, \lambda_n$. The last equality implies that $P^{-1}AP = D$ and hence A is similar to a diagonal matrix.

Fact 6.4

Let A be an $n \times n$ matrix. If A is diagonalizable, then A has n independent eigenvectors.

Proof

Since A is diagonalizable, it follows that there is a nonsingular matrix P such that $D = P^{-1}AP$ where D is a diagonal matrix. Let $v_1, v_2, \ldots, v_n$ be the columns of the matrix P that diagonalizes the matrix A and the diagonal elements of D are $\lambda_1, \lambda_2, \ldots, \lambda_n$. Then, for each, $i = 1, 2, \ldots, n$ we have $Av_i = \lambda_i v_i$. Thus, each v_i, $i = 1, 2, \ldots, n$, is an eigenvector of A. These vectors as the columns of the nonsingular matrix P must be linearly independent. Thus, matrix A has n independent eigenvectors.

Fact 6.5

If A is an $n \times n$ symmetric matrix, then the following statements hold:

1. The eigenvalues of A are all real
2. The eigenvectors corresponding to distinct eigenvalues are orthogonal
3. The matrix A has n independent eigenvectors.

Proof of (2)

Let v_1 and v_2 be two distinct eigenvectors corresponding to two distinct eigenvalues λ_1 and λ_2 of the symmetric matrix A. This implies that $Av_1 = \lambda_1 v_1$ and $Av_2 = \lambda_2 v_2$. Compute the following inner products and use the fact that A is symmetric ($A = A^T$):

$$\lambda_1 < v_1, v_2 >=< \lambda_1 v_1, v_2 >=< Av_1, v_2 >=< v_1, Av_2 >=< v_1, \lambda_2 v_2 >= \lambda_2 < v_1, v_2 >$$

This implies that $(\lambda_1 - \lambda_2) < v_1, v_2 >= 0$. Since λ_1 is not equal to λ_2, it follows that $< v_1, v_2 >= 0$. Therefore, the distinct eigenvectors of the symmetric matrix are orthogonal.

Fact 6.6

For any $n \times n$ real symmetric matrix A, there is an orthogonal matrix P whose columns are the eigenvectors corresponding to the distinct eigenvalues such that $QAP = D$, where D is a diagonal matrix whose diagonal elements are the eigenvalues and Q is the inverse of P.

References

Anton, Howard. *Elementary Linear Algebra*, Seventh Edition, John Wiley & Sons, New York, 1994.

Bauldry, C. William. *Linear Algebra with Maple*, John Wiley & Sons, New York, 1995.

Edelstein-Keshet. *Mathematical Models in Biology*, Random House, New York 1988.

Evans, Benny, and Johnson, Jerry. *Linear Algebra with Derive*, John Wiley & Sons, New York, 1994.

Grossman, I. Stanley. *Elementary Linear Algebra*, Fifth Edition, Saunders, New York, 1994.

Horn A., Roger and Johnson, Charles R. *Matrix Analysis*, Cambridge University Press, Cambridge, 1985.

Johnson, Eugene. *Linear Algebra with Maple*, Brooks/Cole, Pacific Grove, Calif. 1993.

Lay C., David. *Linear Algebra and Its Applications*, Addison–Wesley, Reading, Mass., 1994.

Michael, T.S. The ranks of tournament matrices, *American Mathematical Monthly*, Volume 102, Number 7 (1995), 637–639.

Porter, Gerald J. and Hill, David R., *Introduction to Linear Algebra, A Labratory Course Using MATHCAD*, Springer-Verlag, New York, 1996

Strang, Gilbert. *Linear Algebra and Applications*, Third Edition, Harcourt Brace Jovanovich, San Diego, 1988.

Tucker, Alan. *Linear Algebra, An Introduction to the Theory and Use of Vectors and Matrices*, Macmillan Publishing Company, New York, 1993.

Tucker, Alan. *A Unified Introduction to Linear Algebra: Models, Methods, and Theory*, Macmillan, New York, 1988.

Williams, Gareth. *Linear Algebra with Applications*, Second Edition, Wm. C. Brown, Dubuque, Iowa, 1991.

MapleV and ILAT

MapleV is a computer algebra system capable of performing symbolic, numerical, and graphical computations. MapleV has its own programming environment. Its library has several built-in packages, each of which must be initialized prior to using any of the functions in that package. Using the programming environment of MapleV, you can easily write your own functions and procedures, help files, and libraries. It is a system suited for creating interactive mathematical texts. Of interest to us is the linear algebra package. Before you proceed to perform any computations, enter:

```
> with(linalg);
```

MOST FREQUENTLY USED MAPLE COMMANDS

In this section we design the Maple commands and functions that may arise in an introductory linear algebra course. MapleV has an online help. If you are seeking information on a topic, all you need to enter is a question mark and the filename, for example,

```
> ?solve
```

A Maple input statement must always end with *either a colon or a semicolon*. They both execute the statement. The main difference is that **the semicolon** shows the output while **colon** does not.

Enter a function

A function in Maple is entered as

```
> f:=x-> x^3;
```

Enter a matrix

A 3×4 matrix A is entered as

```
> A:=matrix([[1,3,4,7],[5,6,7,11],[15,1,-1,4]]);
```

If you want to select an **entry**, say the entry in the second row and third column, enter

```
> A[2,3];
```
If you want to select the second **row**, enter
```
> row(A,2);
```
If you want to select the third **column**, enter
```
> col(A,3);
```
If you want to **add** or **multiply** two matrices, first enter the matrices:
```
> A:=matrix([[1,3,4,7] [5,6,7,11] [15,1,-1,4]]);
B:=matrix([[-1,0,1,2] [4,3,1,1] [6,1,1,0]]);
```
To **add**, enter
```
> evalm(A+B);
```
To **multiply**, enter
```
> multiply(A,transpose(B));
```
To **multiply by a scalar**, enter
```
> evalm(4*A);
```
If you want to **swap rows,** say the first and second rows, enter
```
> swaprow(A,1,2);
```
If you want to **multiply a row,** say multiply the third row by 5, enter
```
> mulrow(A,3,5);
```
If you want to **add a multiple of a row to another row,** say multiply row two by 6 and add the result to row three, enter
```
> addrow(A,2,3,6);
```
A 4×1 **column matrix** is
```
> A:=matrix([[1],[2],[3],[-5]]);
```
You may generate, say a 5×7 matrix, via the Maple code
```
> A:=matrix(5,7):
> for i from 1 to 5 do for j from 1 to 7 do
A[i,j]:=i+j;od;od; print(A);
```

Enter a vector

One way of entering a vector is
```
> v:=vector([2,3,5]);
```

If a vector has all components equal to some scalar a, say a vector with five equal components, you enter

```
> v:=vector(5,a);
```

If the components of a vector are given in terms of a function, say x^2, then you enter

```
> f:=x->x^2;
> v:=vector([f(1),f(2),f(3)]);
```

In general you can generate a vector of any dimension, say a vector with five components, using the Maple code

```
> v:=vector(5):
> for i from 1 to 5 do v[i]:=i^2;od:print(v);
```

Solve

First you enter the equations and then you call the solve function:

```
> eq1:=x-y=2; eq2:=2*x-4*y=11;
> solve({eq1,eq2},{x,y});
```

You can use the linear solve to solve the matrix equation Ax = b by entering matrices A and b:

```
> A:=matrix([[1,2],[3,2]]); b:=matrix([[1],[3]]);
> linsolve(A, b, 'r', v);
> A:=matrix([[5,7],[0,0]]): b:=vector([3,0]):
> linsolve(A, b, 'r', v);
```

Plot a function

Suppose you want to plot the function

```
> f:=x->sin(x);
```

You call

```
> plot(f(x),x=0..Pi);
```

The plot function has many options. With it you can specify the range, color, thickness, and style.

```
> plot(f(x),x=0..Pi,y=0..1,color=blue,thickness=10);
```

You may want to display the graph of several functions on the same coordinates. You can do this in one of the following ways:

1. Enter all the functions and use the plot command:

```
> f:=x->x^2; g:=x->exp(x); h:=x->ln(x);
> plot({f(x),g(x),h(x)},x=1...2);
```

2. Enter a sequence of plots, call the package with (plots), and the display the graphs:

```
> p1:=plot(f(x),x=1...2,color=red):
> p2:=plot(g(x),x=1...2,color=blue):
> p3:=plot(h(x),x=1...2,color=green):
> with(plots):
> display({p1,p2,p3});
```

You may be interested in plotting a relation in x and y such as

```
> x^2+y^2=1;
> implicitplot(x^2+y^2=1,x=-1...1,y=-1...1);
```

The three-dimensional version is

```
> implicitplot3d(x^2+y^2+z^2=1,x=-1...1,y=-1...1,z=0..1);
```

ILAT PACKAGES

We have added to the MapleV library six ILAT packages: linsys, linmat, linspace, linpdt, lintran, and lineign. Each package consists of many functions and each function has three modes: **Demostration**, **Interactive**, and **Nostep**. On executing a function, you select the desired mode.

If MapleV has a similar function, we have used the same name. Before proceeding to perform any computations, you must initialize the package you need along with linalg. For example, if you want to use linsys, then you need to intialize

```
> with(linalg): with(linsys);
```

linsys

This package includes the functions gausselim, rref, graph, solveqns, and backsub.

To call gausselim, enter the matrix and then the function

```
> A:=matrix([[1,2,3],[3,2,1]]);
> gausselim(A);
```

To call rref, enter the matrix and then the function

```
> A:=matrix([[1,2,3],[3,2,1],[4,7,9]]);
> rref(A);
```

To call graph, enter the equations and then the function

```
> eq1:=x+2*y=1; eq2:=-x+3*y=2;
> graph(eq1,eq2);
```

To call backsub, enter the matrix in echelon form and then the function

```
> A:=matrix([[1,2,3],[0,1,1],[0,0,0]]);
> backsub(A);
```

To call `solveqns`, enter the equations and then the function

```
> eq1:=x+2*y=1; eq2:=-x+3*y=2;
> solveqns({eq1,eq2},{x,y});
```

All the above functions can be called without an argument. For example, you may call

```
> gausselim();
```

and press enter, then enter the inputs. This is true for all functions in our packages.

linmat

This package includes the functions `inverse`, `LUdecomp`, `Geometry`, `matrixmul`, `commute`, `trsum`, `trproduct`, `transtrans`, and `trinverse`.
To call `inverse`, enter the matrix and then the function

```
> A:=matrix([[1,2,3],[3,2,1]]);
> inverse(A);
```

To call `LUdecomp`, enter the matrix and then the function

```
> A:=matrix([[1,2,3],[3,2,1]]);
> LUdecomp(A);
```

To call `Geometry`, enter the matrix and the set of points and then the function

```
> S:={[1,2],[1,3],2,3]}:A:=matrix([[1,-1],[1,0]]);
> Geometry(S,A);
```

To call `matrixmul`, enter the matrices and then the function

```
> A:=matrix([[1,2,3],[3,2,1]]);B:=matrix([[1],[3],[5]]);
> matrixmul(A,B);
```

To call `commute`, enter the matrices and then the function

```
> A:=matrix([[1,2],[2,1]]);B:=matrix([[1,5],[6,7]]);
> commute(A,B);
```

To call `trsum`, enter the matrices and then the function

```
> A:=matrix([[1,5],[7,1]]);B:=matrix([[1,9],[6,7]]);
> trsum(A,B);
```

To call `trproduct`, enter the matrices and then the function

```
> A:=matrix([[1,5],[7,1]]);B:=matrix([[1,9],[6,7]]);
> trproduct(A,B);
```

To call `trinverse`, enter the matrix and then the function

```
> A:=matrix([[1,5],[7,1]]);
> trinverse(A);
```

linspace

This package includes the functions lincomb, lindep, subspace, graphvectadd, graphscalarmulti, graphlincomb, and basis.
To call lincomb, enter the vector to be expressed as a combination of a set of vectors and then the function

```
> w:= vector([6,7,9]); v1:=vector([1,2,3]);
v2:=vector([4,-1,1]); v3:=vector([-1,0,1]);
> lincomb(v1,v2,v3,w);
```

To call lindep, enter the set of vectors and then the function

```
> v1:=vector([1,2,3]); v2:=vector([4,-1,1]);
v3:=vector([-1,0,1]);
> lindep(v1,v2,v3);
```

To call subspace, enter the set S and then the function

```
> S:={(x,y),x+y=0};
> subspace(S);
```

To call basis, enter the set of vectors and then the function

```
> v1:=vector([1,2,3]); v2:=vector([4,-1,1]);
v3:=vector([-1,0,1]);
> basis(v1,v2,v3);
```

linpdt

This package includes the functions GramSchmidt, QRdecomp, lsqrdemo, and leastsqrs.
To call GramSchmidt, enter the set of vectors and then the function

```
> v1:=vector([1,2,3]); v2:=vector([4,-1,1]);
v3:=vector([-1,0,1]);
> GramSchmidt(v1,v2,v3);
```

To call QRdecomp, enter the matrix and then the function

```
> A:=matrix([[1,2,3],[4,-1,1],[-1,0,1]]);
> QRdecomp(A);
```

To call leastsqrs, enter the set of data points or the augmented matrix of the underlying system of equation and then the function:

```
> A:=matrix([[1,2,3],[4,-1,1],[-1,0,1],[1,1,-4]]);
> leastsqrs(A);
```

lintran

This package includes the functions `lineartran`, `kernel`, `range`, `matrixrep`, `changebasis`, and `BaseGeometry`.

To call `lineartran`, enter the transformation and then enter the function whose arguments are the transformation T, the domain R^3, and the range R^2

```
> T:=x->(x[1]-x[2],x[2]-x[3]);
> lineartran(T,R3,R2);
```

To call `kernel`, enter the transformation and then enter the function whose arguments are the transformation T, the domain R^4, and the range R^3

```
> T:=x->(x[1]-x[2],x[2]-x[3],x[4]-2*x[1]);
> kernel(T,R4,R3);
```

To call `range`, enter the transformation and then enter the function whose arguments are the transformation T, the domain R^4, and the range R^3

```
> T:=x->(x[1]-x[2],x[2]-x[3],x[4]-2*x[1]);
> range(T,R4,R3);
```

To call `matrixrep`, enter the transformation and then enter the function whose arguments are the transformation T, the domain R^3, and the range R^2

```
> T:=x->(x[1]-x[2],x[2]-x[3]);
> matrixrep(T,R3,R2);
```

To call `changebasis`, enter the matrices whose columns are the given bases and then enter the function:

```
> G:=matrix([[1,2],[2,1]]); H:=matrix([[1,2],[2,5]]);
> changebasis(T,R4,R3);
```

lineign

This package includes the functions `eigenvals`, `eigenvects`, `diagonalize`, and `SVdecomp`.

To call `eigenvals`, enter the matrix and then the function

```
> A:=matrix([[1,2,3],[3,2,1],[1,0,1]]);
> eigenvals(A);
```

To call `eigenvects`, enter the matrix and then the function

```
> A:=matrix([[1,2,3],[3,2,1],[1,0,1]]);
> eigenvects(A);
```

To call `diagonalize`, enter the matrix and then the function

```
> A:=matrix([[1,2,3],[3,2,1],[1,0,1]]);
> diagonalize(A);
```

To call SVdecomp, enter the matrix and then the function

```
> A:=matrix([[1,2,3],[3,2,1],[1,0,1]]);
> SVdecomp(A);
```

USING MAPLE FUNCTIONS INSTEAD OF ILAT FUNCTIONS

MapleV and ILAT functions share the same name whenever both functions are available in MapleV and ILAT. If you want to use functions from the linalg package in MapleV instead of ILAT functions, enter gausselim as

```
> linalg[gausselim](A);
```

In general, you enter

```
> linalg[function-name](arguments);
```

Index